CALVERT
MATH

Practice and Enrichment Workbook

Calvert Math is based upon a previously published textbook series. Calvert School has cus-
tomized the textbooks using the mathematical principles developed by the original authors.
Calvert School wishes to thank the authors for their cooperation. They are:

Audrey V. Buffington
Mathematics Teacher
Wayland Public Schools
Wayland, Massachusetts

Alice R. Garr
Mathematics Department Chairperson
Herricks Middle School
Albertson, New York

Jay Graening
Professor of Mathematics
    and Secondary Education
University of Arkansas
Fayetteville, Arkansas

Philip P. Halloran
Professor,
Mathematical Sciences
Central Connecticut State University
New Britain, Connecticut

Michael Mahaffey
Associate Professor,
    Mathematics Education
University of Georgia
Athens, Georgia

Mary A. O'Neal
Mathematics Laboratory Teacher
Brentwood Unified Science
    Magnet School
Los Angeles, California

John H. Stoeckinger
Mathematics Department Chairperson
Carmel High School
Carmel, Indiana

Glen Vannatta
Former Mathematics Supervisor
Special Mathematics Consultant
Indianapolis Public Schools
Indianapolis, Indiana

WORKBOOK AUTHORS

Shannon Mullineaux   Victoria T. Strand

COPY EDITORS

Bernadette Burger, Senior Editor
Sarah E. Hedges
Maria R. Kerner
Mary Pfeiffer
Megan L. Snyder

GRAPHIC DESIGNERS

Vickie M. Johnson, Senior Designer
Vanessa Ann Panzarino

PROJECT FACILITATOR/LEAD RESEARCHER

Nicole M. Henry, Math Curriculum Specialist

SENIOR CONSULTANT/PROJECT COORDINATOR

Jessie C. Sweeley, Manager of Curriculum Development

ISBN-13: 978-1-888287-56-1

1 2 3 4 5 6 7 8 9 10   12 11 10 09 08

Contents

Name _____

© Calvert School

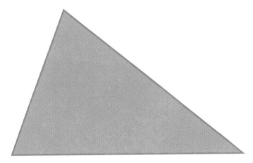  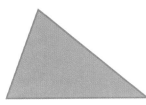

Mark ✔ on the largest object. Circle the smallest object.

Chapter 1

Name

 Circle the thicker sandwich.

 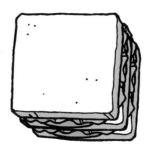

 Circle the thinner towel.

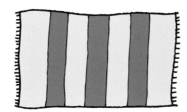 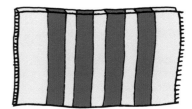

 Circle the thicker button.

 Circle the thinner piece of watermelon.

Name

 Color a purple dot under the child who is farthest from the house.

○   ○   ○

 Color a brown dot under the child who is playing farthest from the tree.

○   ○   ○

 Color a red dot under the child who is the nearest to the garden.

○   ○   ○   ○   ○

Name _____

Draw a doll inside the toy box. Draw a ball outside the toy box.

Chapter 1

Name

Color the bird flying over the tree blue. Color the chipmunk on the tree brown.
Color the rabbit under the table red.

Chapter 1

(five) **5**

Name _____

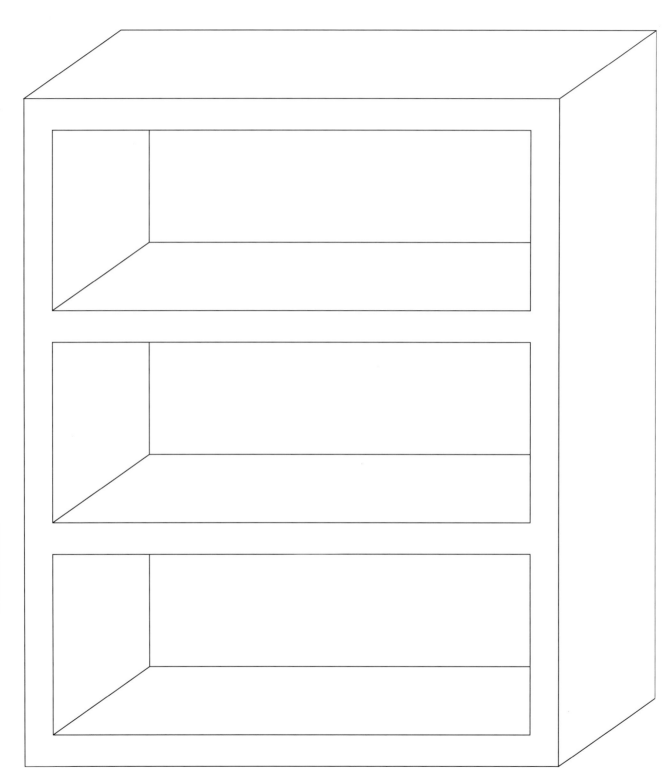

Draw a flower on the middle shelf. Draw a smiley face on the bottom shelf. Draw a sun on the top shelf.

Chapter 1

Name

Draw a tree to the left of the house.
Draw a ball on the right side of the path.
Draw a flower in the middle of each window box.
Draw bushes on the left side of the path.

Name

Problem Solving

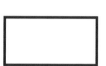 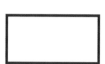 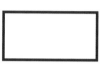 

Circle the object that is different.

Name

Cut out the candy and sort by color.
Fill in the chart on the following page.

(nine) 9

Name _____

Color	How many?
[ ] brown	
[ ] red	
[ ] purple	
[ ] green	
[ ] orange	
[ ] blue	

Color the boxes in the first column. Count how many of each color of candy you have. Write the number of each candy in the third column.

Name

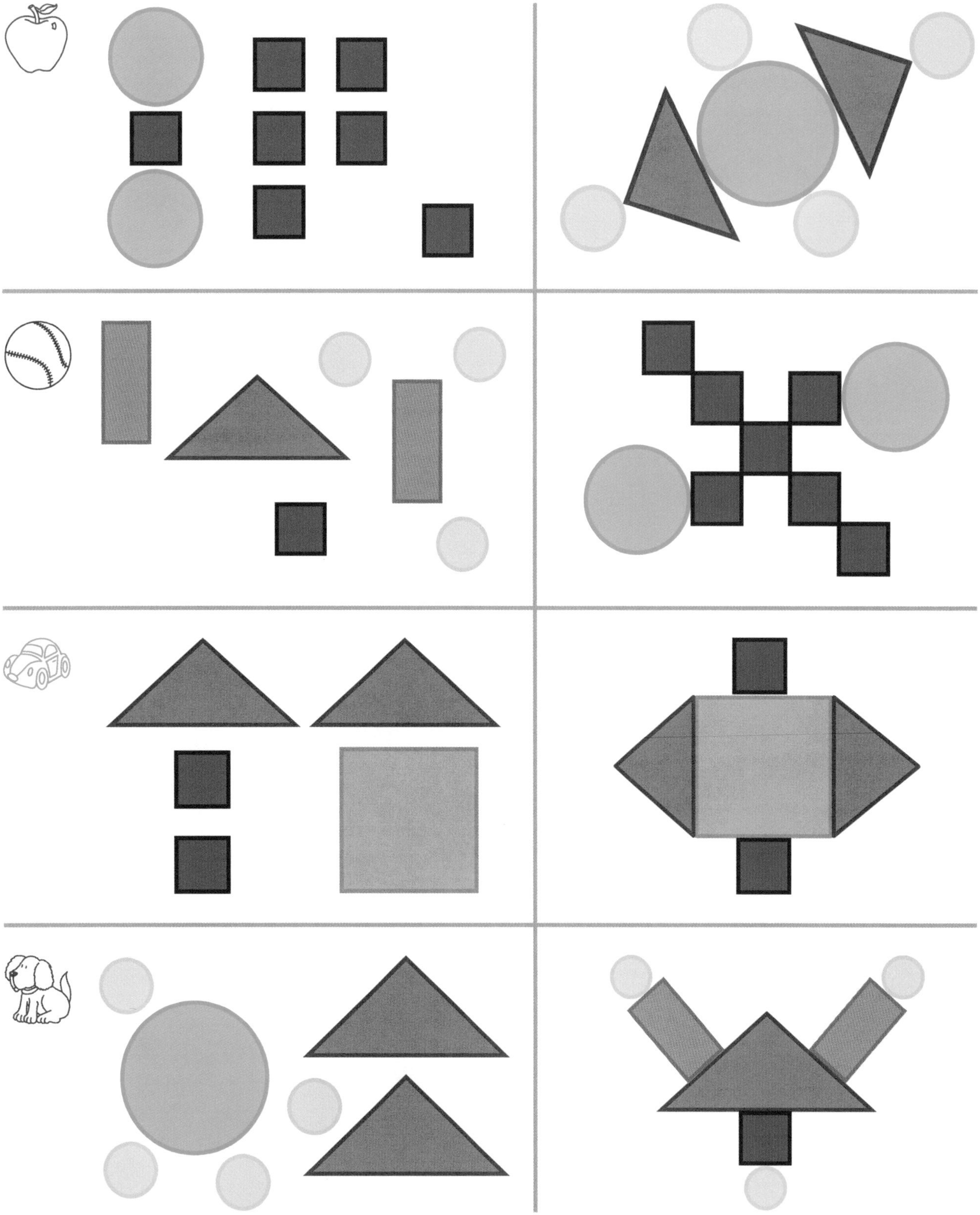

Match the pieces on the left with their design on the right.

Chapter 1

Name

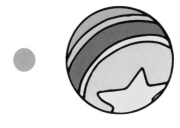

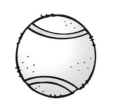

Draw a line to connect the objects that are the same size.

Name _____

Cut out and put in order of size, largest to smallest.

Name

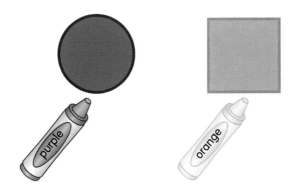

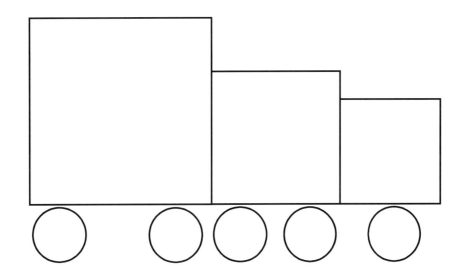

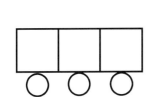 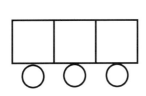 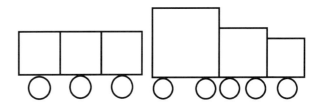

Color the shapes as shown.

© Calvert School

Name

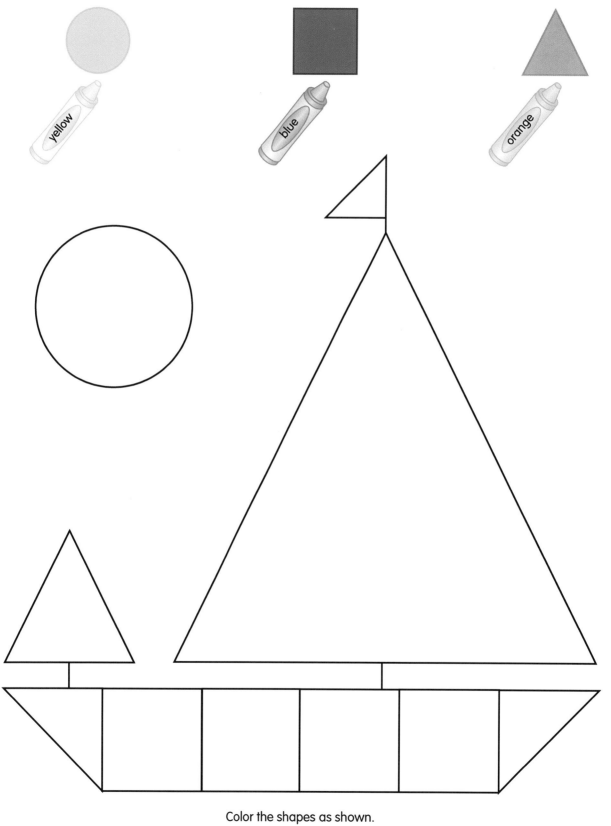

Color the shapes as shown.

18 (eighteen)

Chapter 2

Name

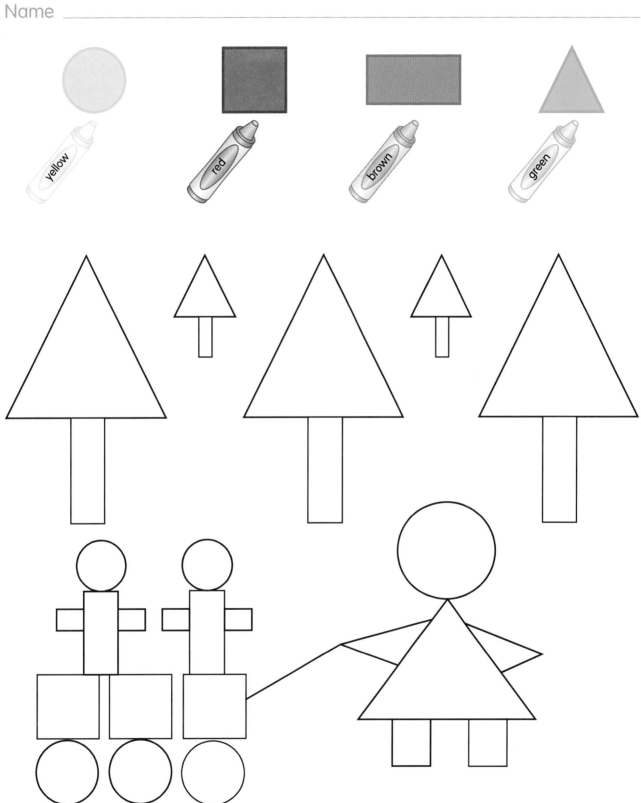

Color the shapes.

Chapter 2

Name

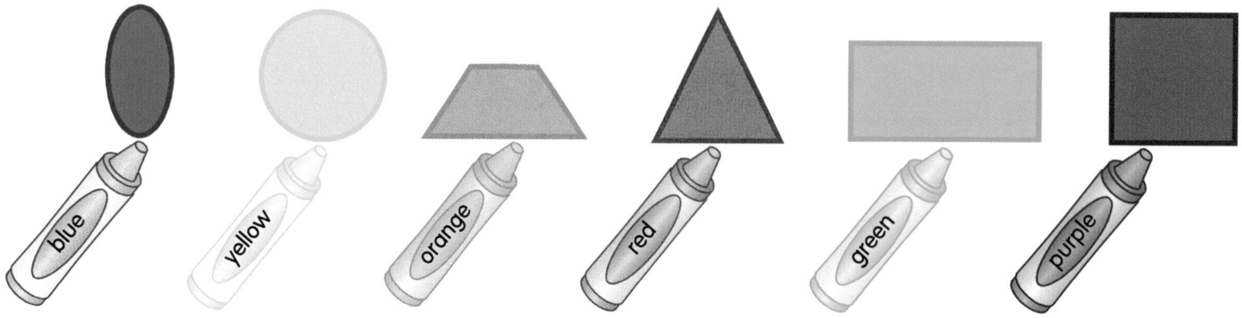

blue   yellow   orange   red   green   purple

Color the shapes.

Create a picture of your choice out of shapes.

My
Shape
Book

Create an animal out of shapes.

Square

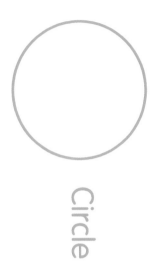

Circle

Triangle

Heart

Diamond

Trapezoid

Polygon

Rectangle

Oval

Name

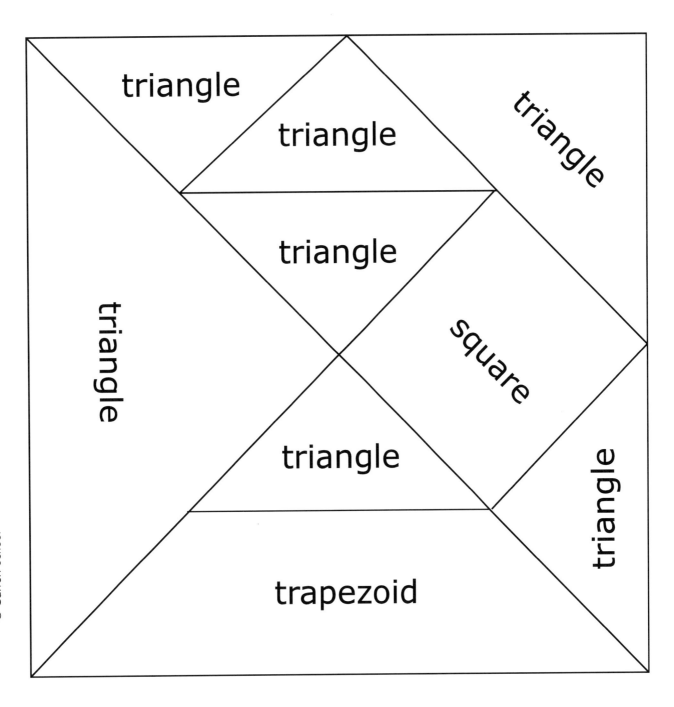

triangle

triangle

triangle

triangle

triangle

triangle

square

triangle

trapezoid

Cut out tangram puzzle shapes.
Discuss each shape.
Can the child make one large square again?

Name

 Circle the block that comes next when you flip it.

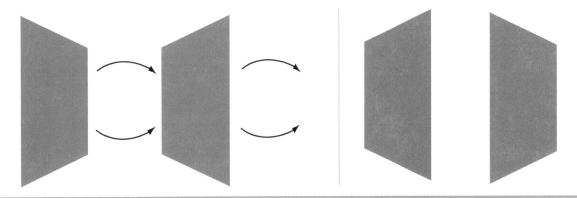

 Circle the block that comes next when you turn it.

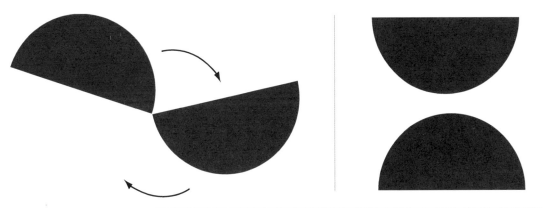

 Circle the block that comes next when you slide it.

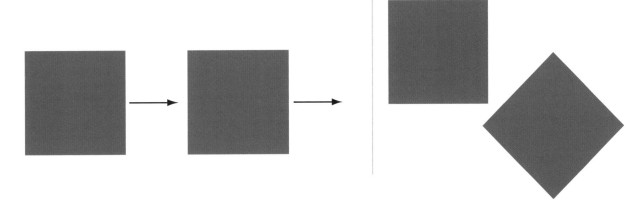

Name _____

 Circle the objects shaped like spheres.

 Draw 2 objects shaped like a sphere.

Name

 Circle the objects shaped like cubes.

 Draw 2 objects shaped like a cube.

Name _____

 Put an X on the cylinders.

 Draw 2 objects shaped like a cylinder.

Name

 Put an X on the cones.

 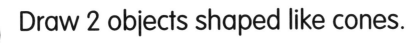 Draw 2 objects shaped like cones.

Name

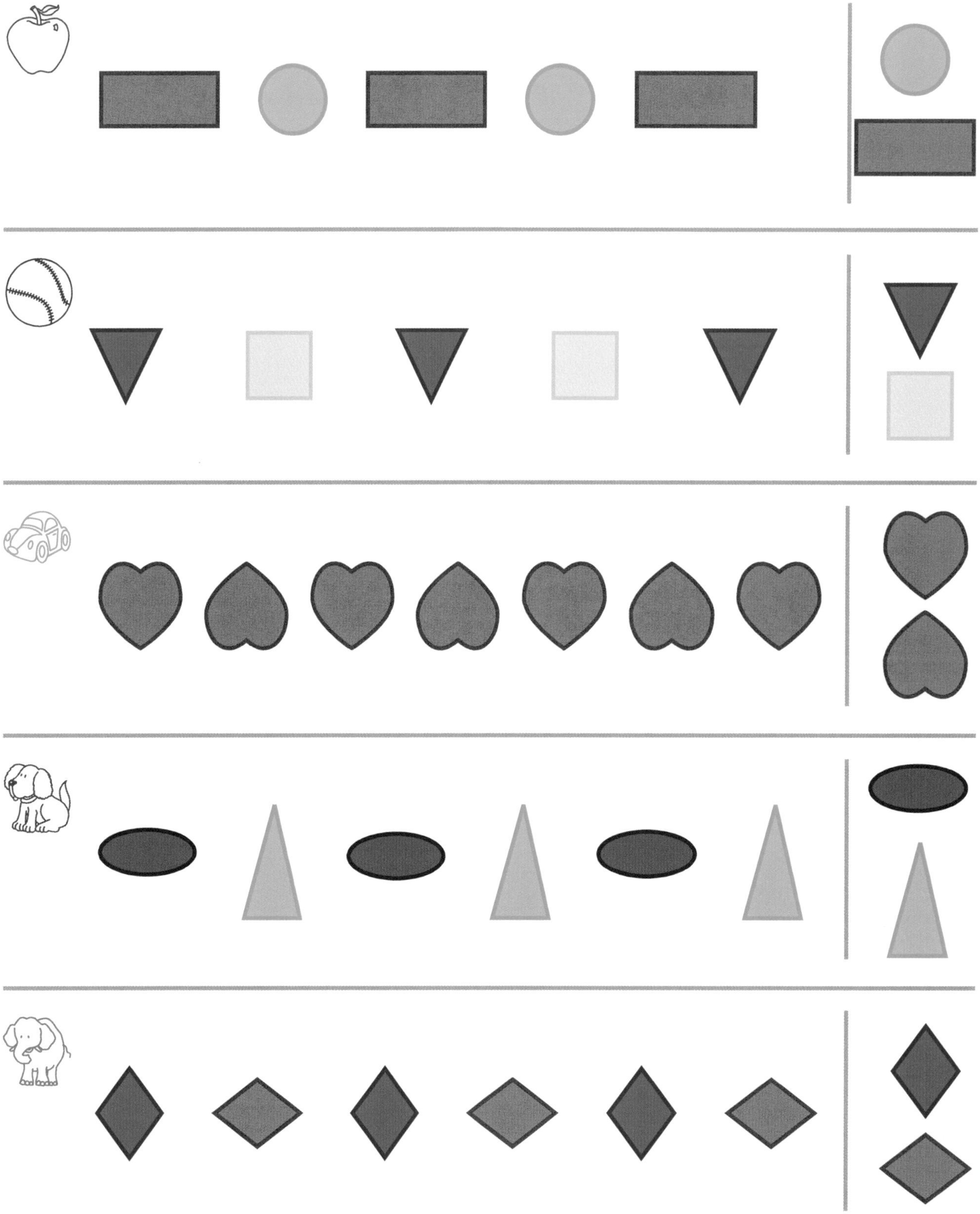

Circle the shape in each row that completes the pattern.

Name

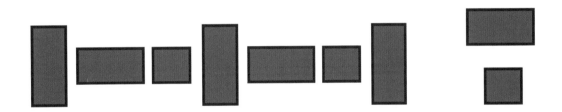

Circle the shape in each row that completes the pattern.

Name _____

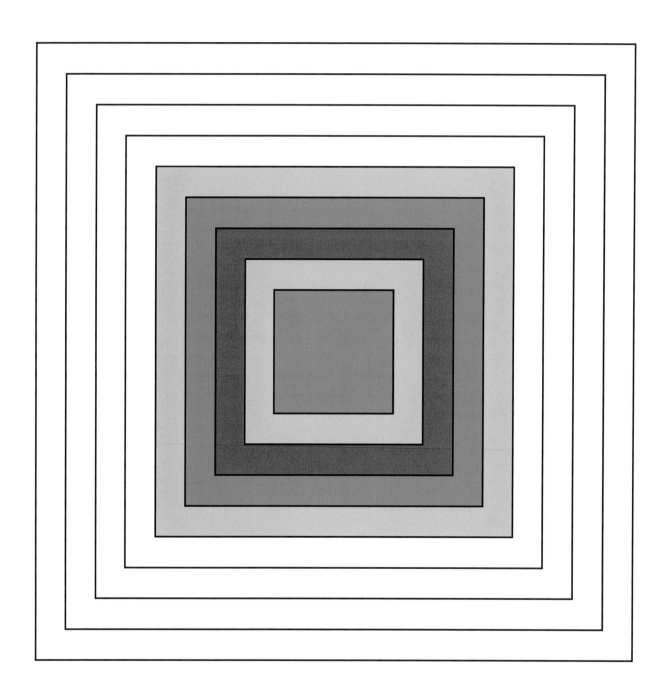

Continue the pattern.

Name _____

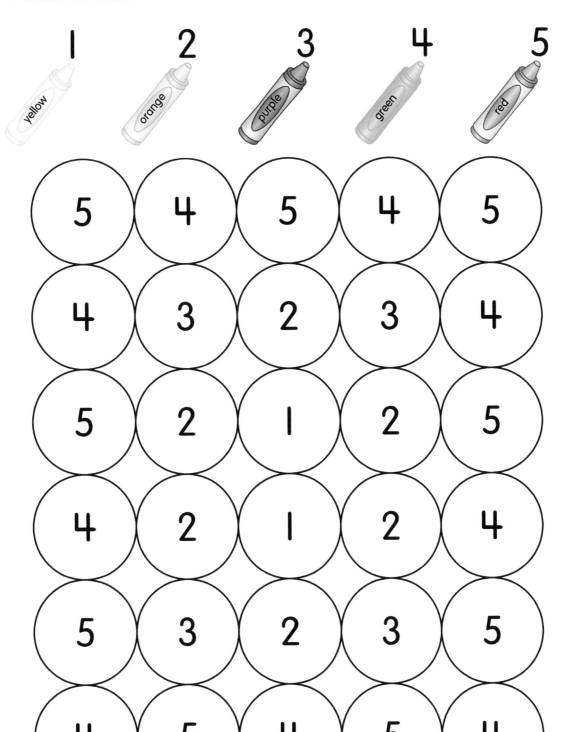

Color the numbers to create a pattern.

Name _____

Choose three colors and make a repeating pattern.

Name _____

_____

_____

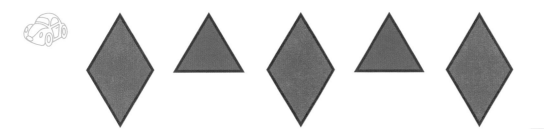

_____

_____

 Make your own 2-shape pattern.

____ ____ ____ ____ ____

Draw the next shape to complete each pattern.

Name _____

 This is an ABC pattern.

A   B   C   A   B   C   A B C

 Name these patterns.

A   A   B   A   A   B

A   A   B   B   ___ ___ ___ ___ ___ ___

___ ___ ___ ___ ___ ___ ___ ___

Name _____

 Draw one candle on each cake.

 Draw one ball next to each bat.

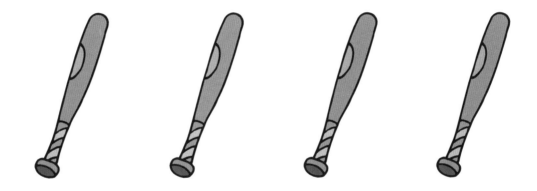

 Draw one bird in each nest.

Name _____

 ⚪ ⚪

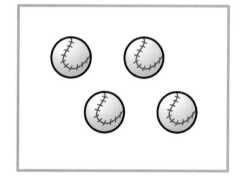 ⚪ ⚪

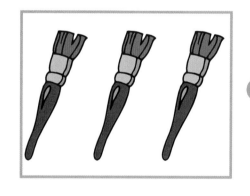 ⚪ ⚪

 ⚪ ⚪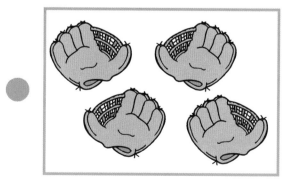

Match groups having the same number.

Name

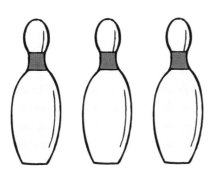

Draw circles to show the same number.

Name _____

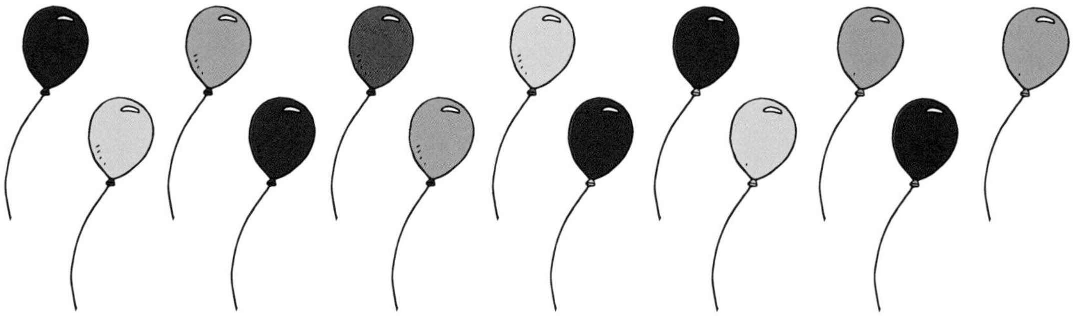

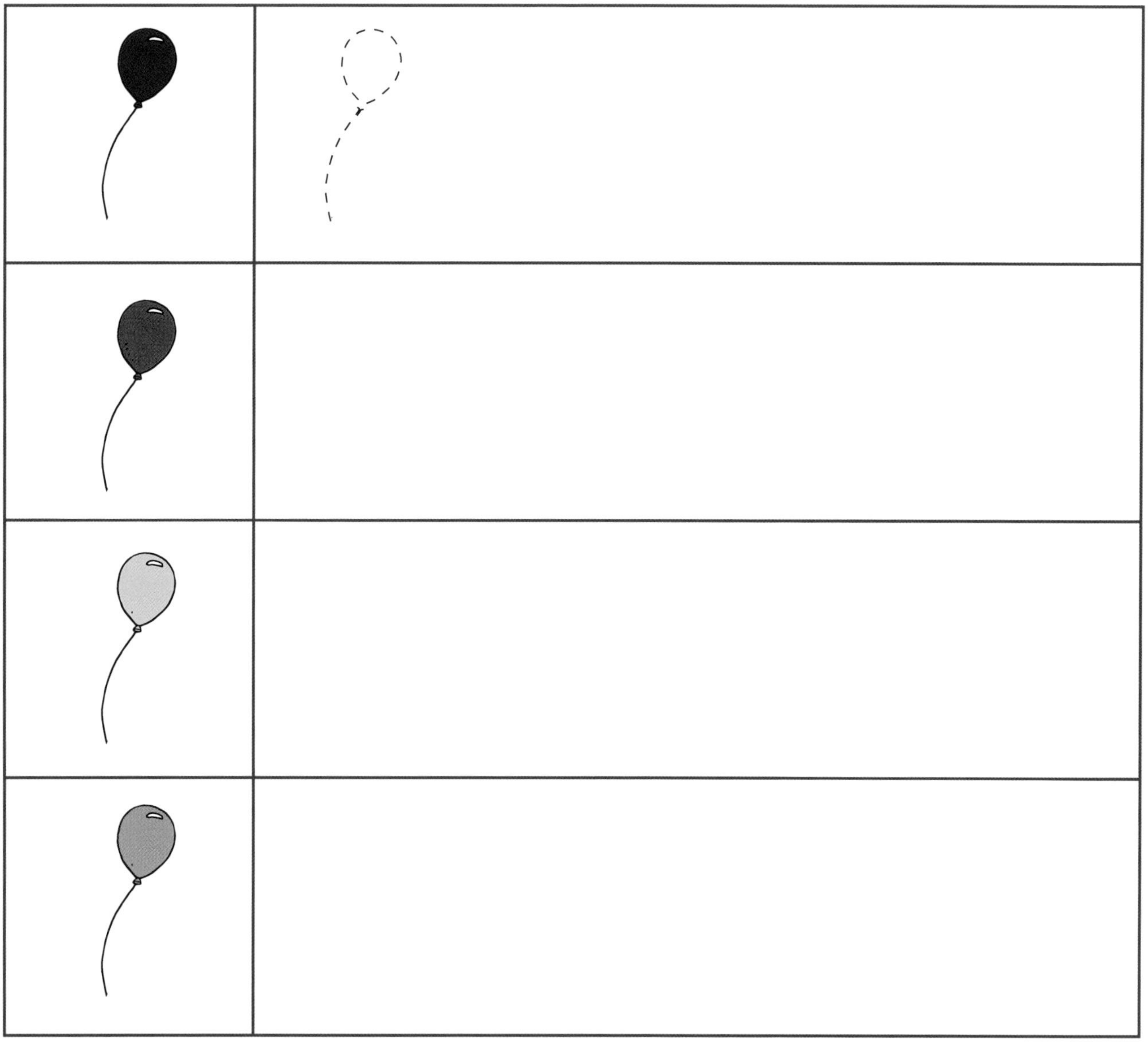

Draw the number of balloons with the same color.

Name _____

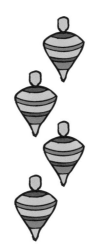

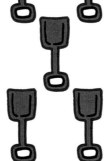

Circle the group in each box that has more.

Chapter 3

Name

Draw a group with one more.

Chapter 3

Name _____

 |

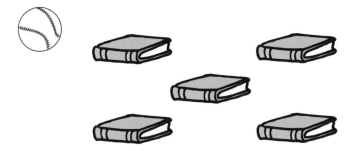 | 

  |

 |

Circle the box in each row with one more.

Chapter 3

Name _____

red	green	blue
▲▲▲		
▲ ▲		
▲▲ ▲▲▲		
▲		

Draw a set of green triangles that has one more than the red.
Then, draw a set of blue triangles that has one more than the green.

Chapter 3

Name

 Put a check on the branch with the most owls.

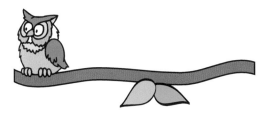

 Put a check on the branch with the most birds.

Name

Draw a set with one less.

Chapter 3

Name _____

Circle the set in each row with one less.

Chapter 3

Name _____

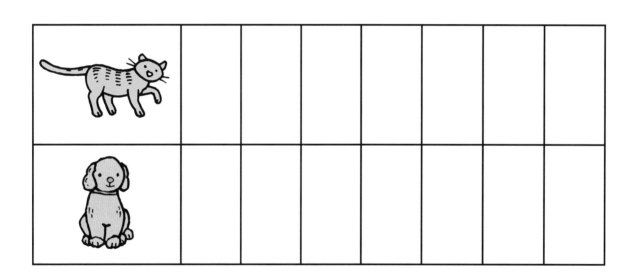

Look at the picture at the top. Count all the cats. Color one rectangle in
the graph for each cat. Count the dogs. Color one rectangle in the
graph for each dog. Circle the animal that has one more.

Chapter 3

Name

 Circle the log that has the least number of frogs.

 Circle the pond that has the least number of ducks.

© Calvert School

Name

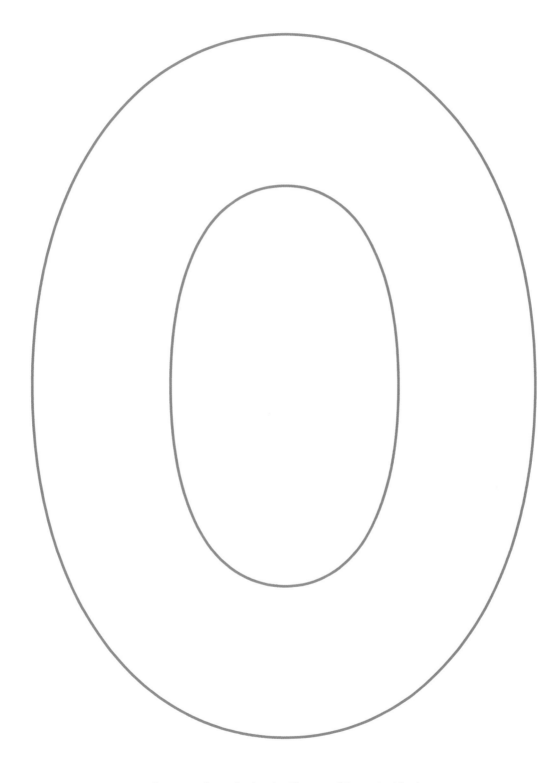

Remove this page from the book. Glue small items inside the zero
(pieces of paper, buttons, sequins, beans, noodles, rice, sand)
to make the zero shape.

Name

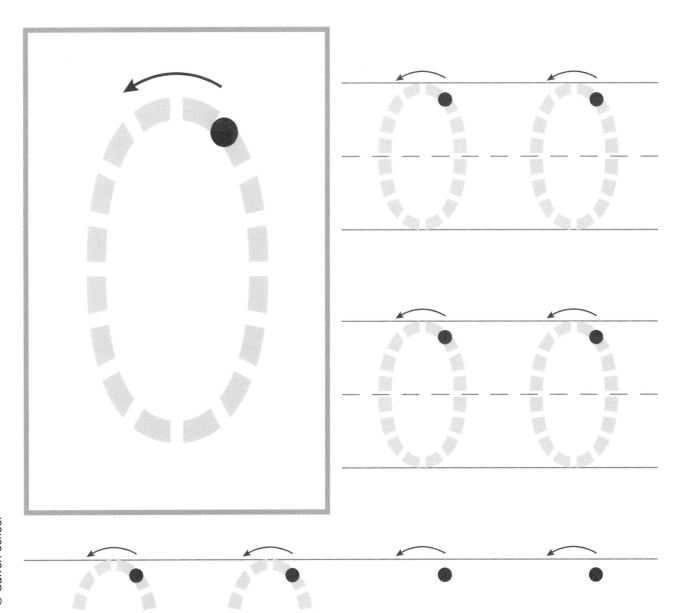

Write 0. Remember to start at the top, draw left, and follow the arrow.

Name _____

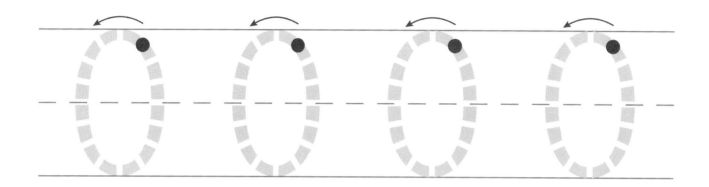

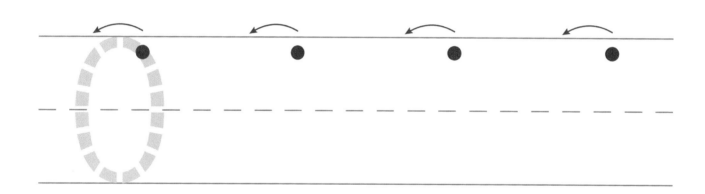

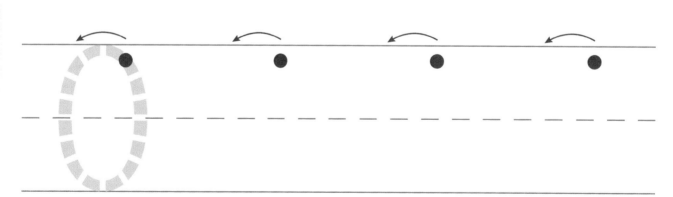

Write a 0. Follow the arrows.

Chapter 4

Name _____

Color the leaf. Cut out the bees. Glue 0 bees on the leaf.

Name _____

Remove this page from the book. Glue small items inside the one
(pieces of paper, buttons, sequins, beans, noodles, rice, sand) to make
the one shape.

Name _____

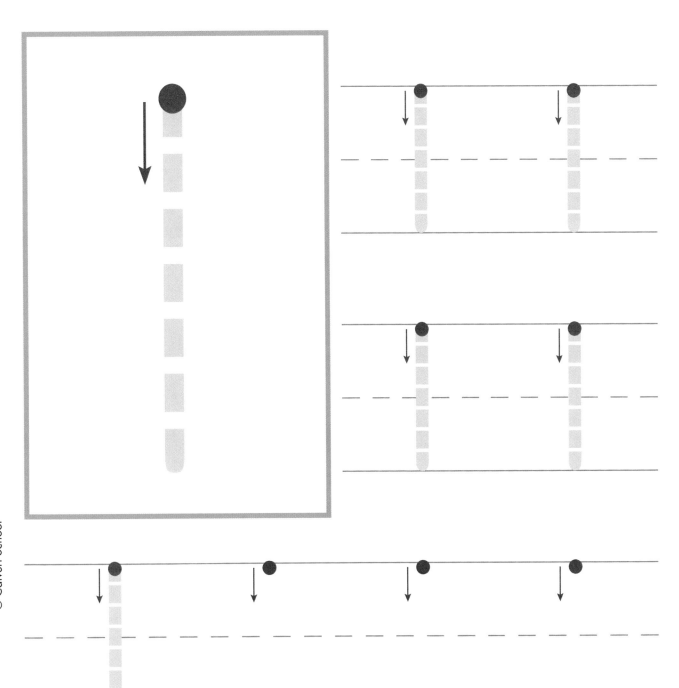

Write 1. Start at the top and draw a line down.

Chapter 4

Name _____

Draw 1 △.

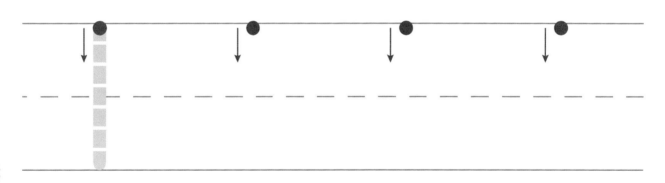

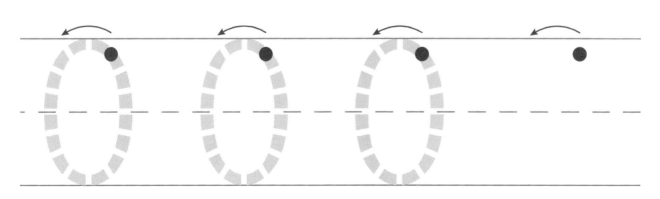

Write 1 and 0. Follow the arrow.

Chapter 4

Name _____

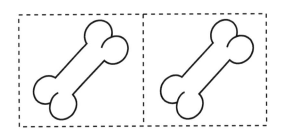

Color the dog. Cut out the bones. Give him 1 bone to eat.
Glue it on the page.

Name _____

Remove this page from the book. Glue small items inside the two (pieces of paper, buttons, sequins, beans, noodles, rice, sand) to make the two shape.

Name _____

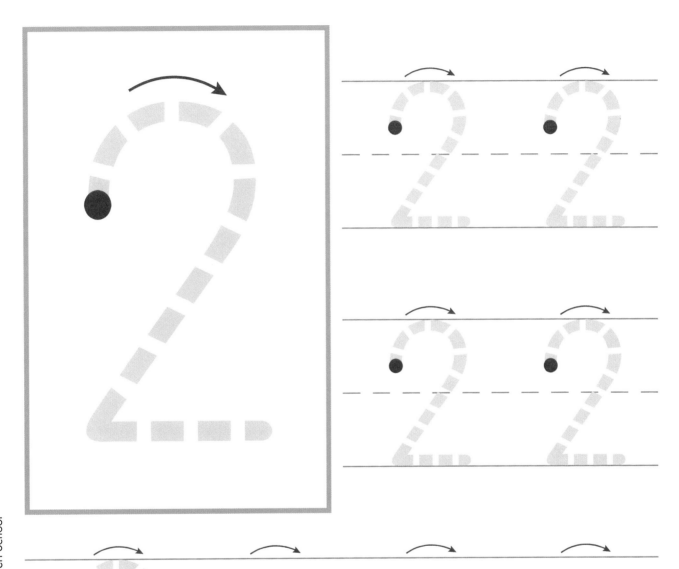

Write 2. Follow the arrow.

Name

Draw 2 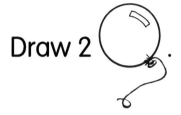.

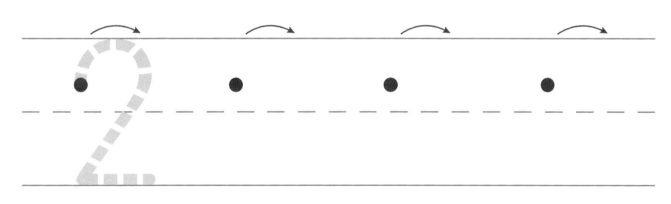

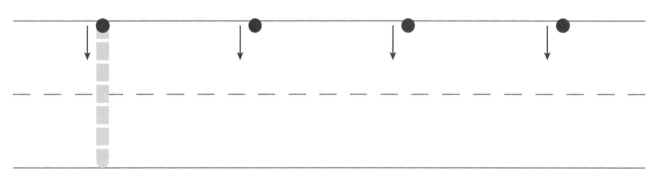

Write 2 and 1.

Name _____

 Draw 1   .

Draw 1  .

 Draw 2  .

Draw 2 .

Name

Find each matching pair of mittens. Color each matching pair with the same colors.

Chapter 4

Name

Color the basket. Cut out the strawberries. Glue 2 strawberries inside the basket.

Name _____

Remove this page from the book. Glue small items inside the three
(pieces of paper, buttons, sequins, beans, noodles, rice, sand) to make
the three shape.

Name _____

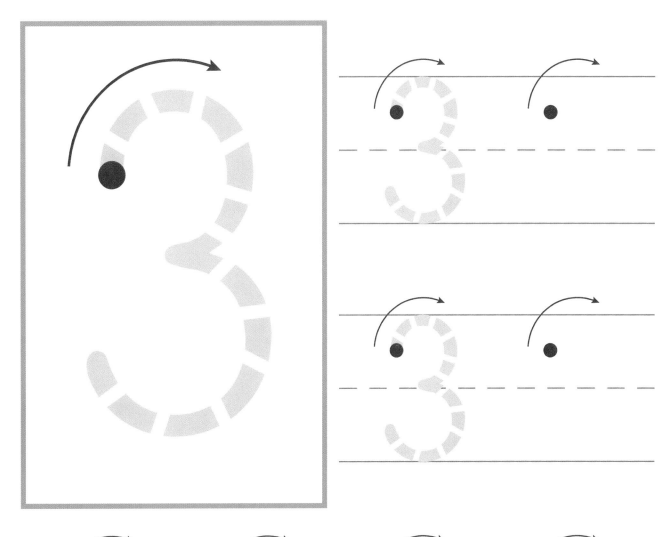

Write 3. Follow the arrow.

Chapter 4

Name _____

Draw 3 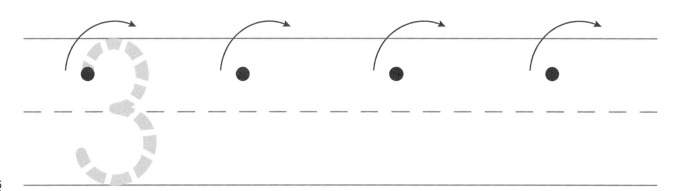.

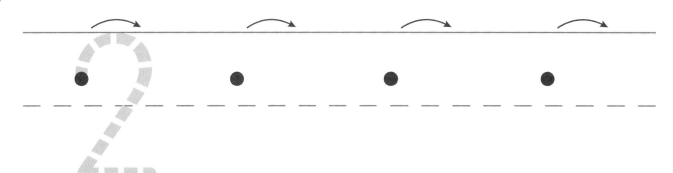

Write 3 and 2.

Name _____

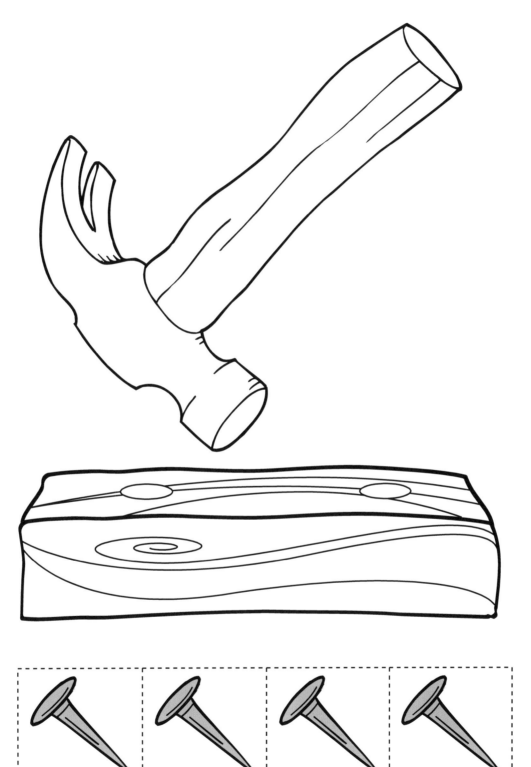

Color the hammer and wood. Cut out the nails.
Glue 3 nails onto the wood.

Chapter 4

(seventy-nine) **79**

Name

 Draw 2   .

Draw 2   .

Draw 3  .

Draw 3 .

Name

Draw 3 colorful buttons on each clown.

Chapter 4

Name

Remove this page from the book. Glue small items inside the four (pieces of paper, buttons, sequins, beans, noodles, rice, sand) to make the four shape.

Chapter 4

(eighty-three) **83**

Name _____

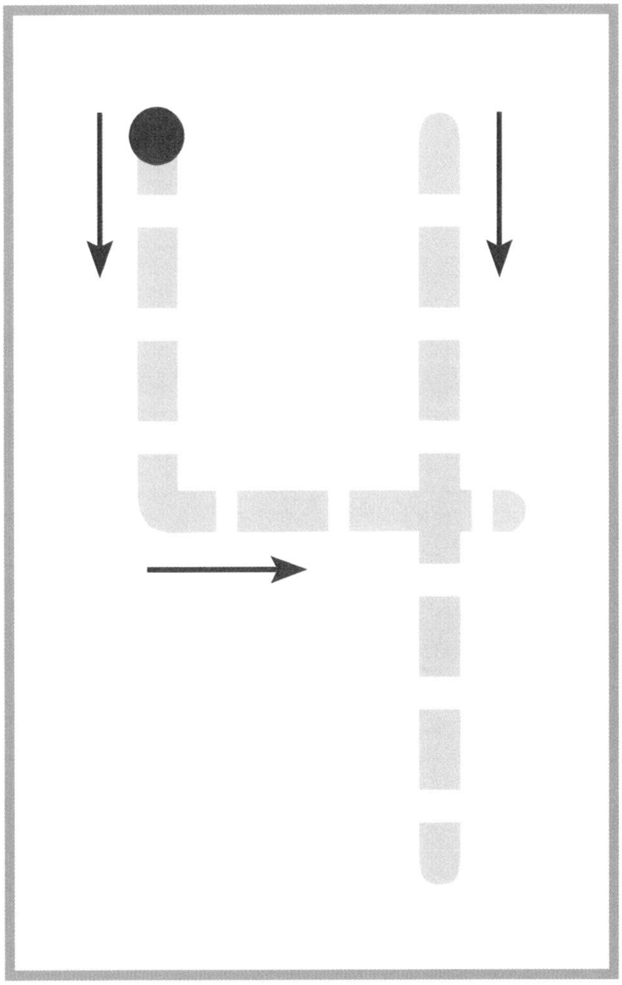

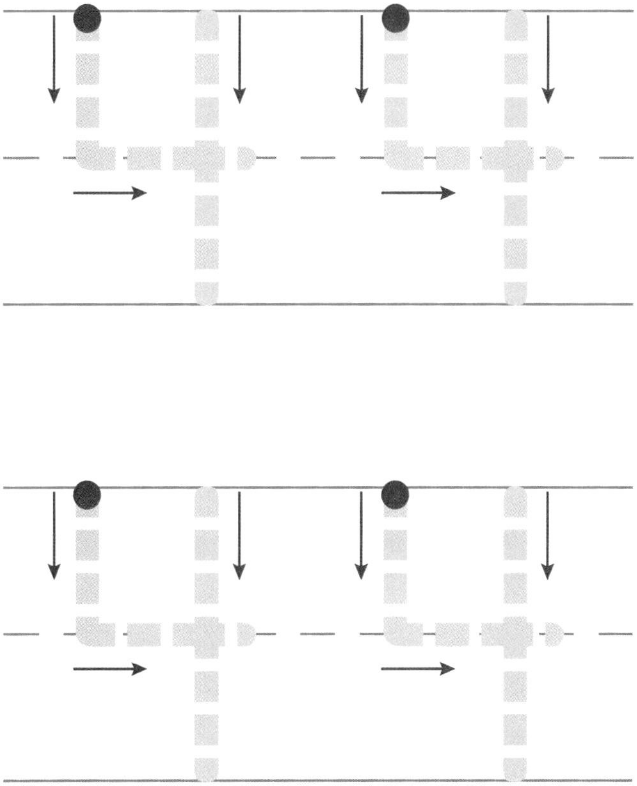

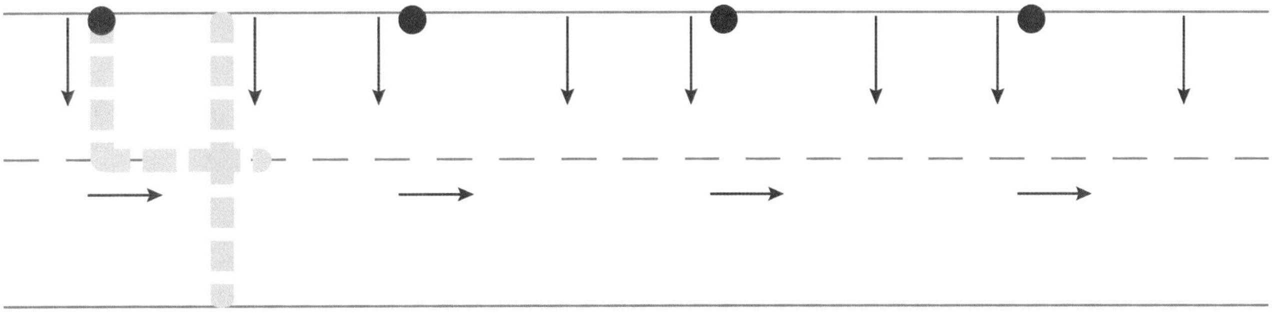

© Calvert School

Write 4. Start with the red dot and follow the arrows.

Name _____

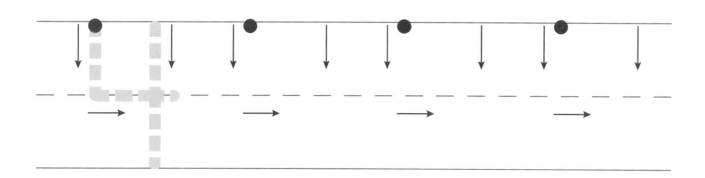

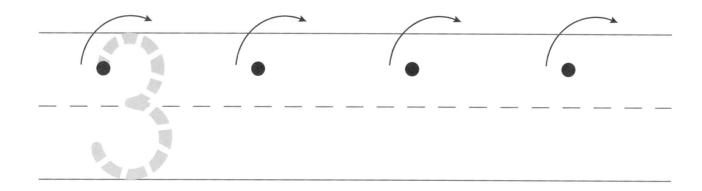

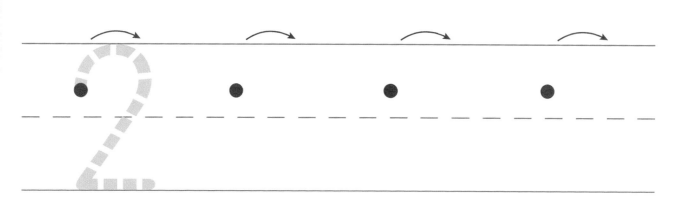

Write 4, 3, and 2.

Name _____

Color the vase. Draw 4 flowers in the vase.

Name _____

 Draw 3 🛥️ .

Draw 3 🍎 .

 Draw 4 🎳 .

Draw 4 🍌 .

Name _____

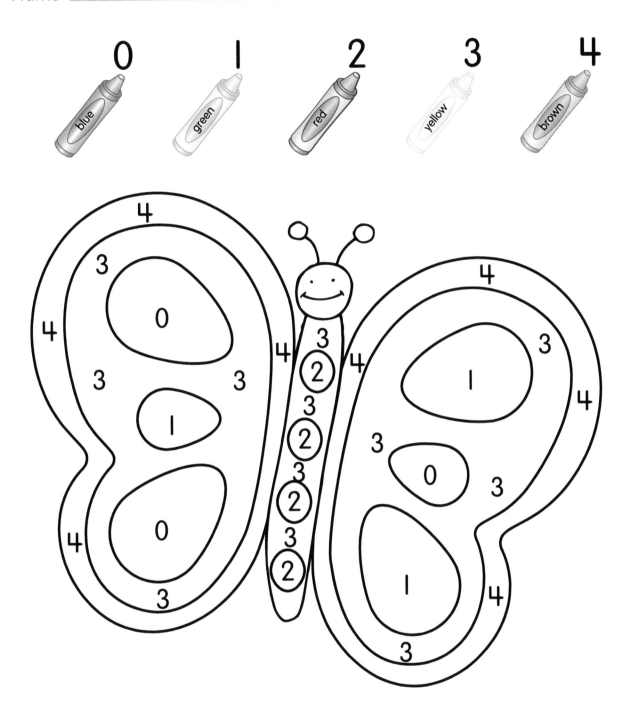

Color the butterfly.

Name _____

Remove this page from the book. Glue small items inside the five
(pieces of paper, buttons, sequins, beans, noodles, rice, sand) to make
the five shape.

Name _____

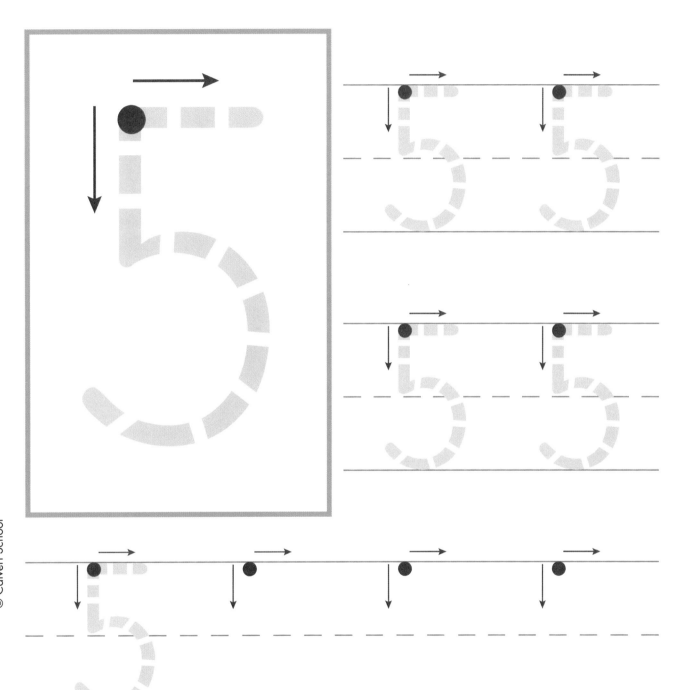

Write 5. Follow the arrows.

Chapter 4

(ninety-three) **93**

Name

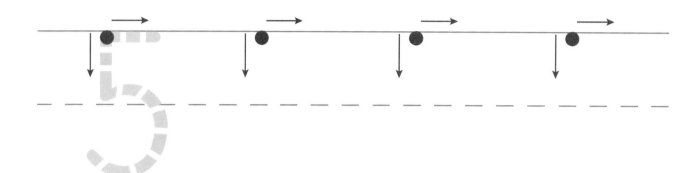

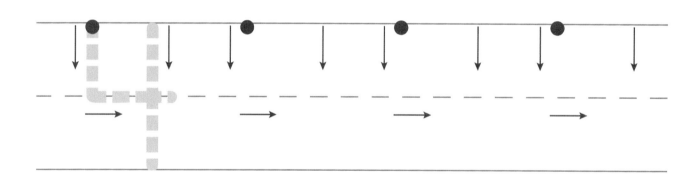

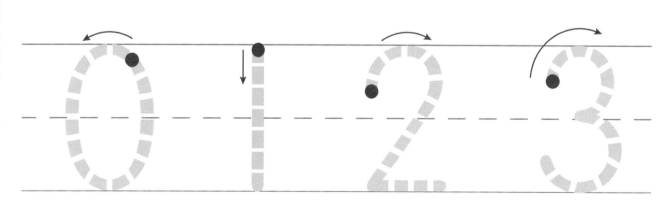

Write the numbers.

Name _____

Draw 5 oranges on the tree. Color the tree.

Chapter 4

Name _____

 Draw 4 🖍 .

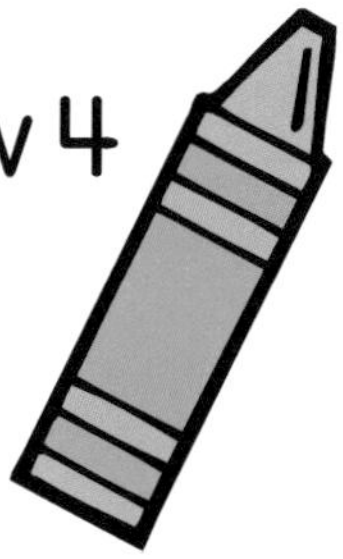

Draw 4 ⚾ .

 Draw 5  .

Draw 5  .

Name _____

0	
5	
3	
1	
2	
4	

Color to show the number.

Name

Remove this page from the book. Glue small items inside the six (pieces of paper, buttons, sequins, beans, noodles, rice, sand) to make the six shape.

Name _____

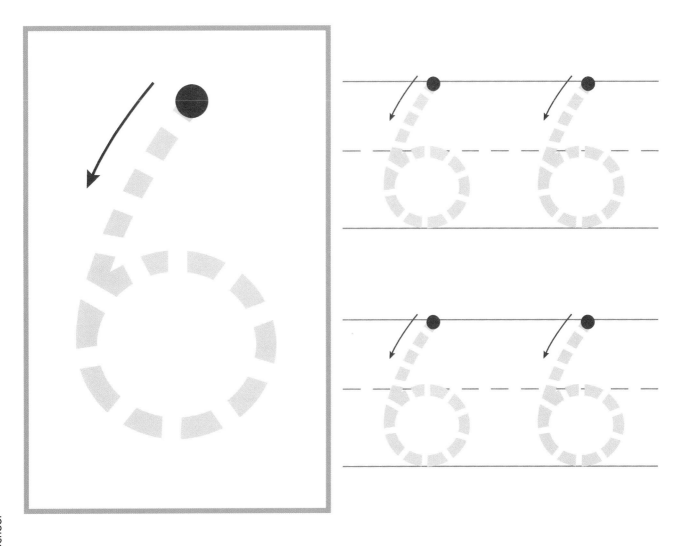

Write 6. Start at the top and follow the arrow.

Chapter 5

Name _____

Color 6 .

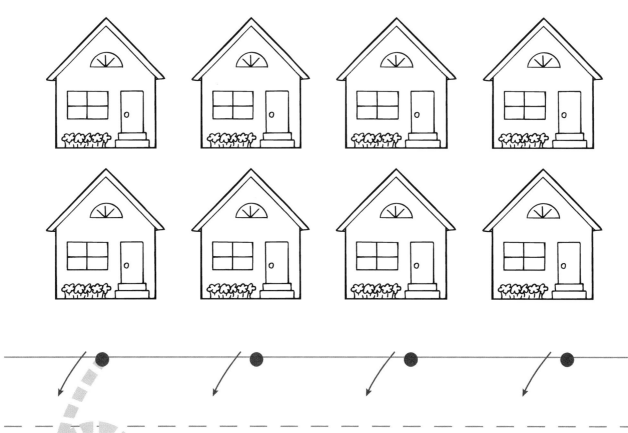

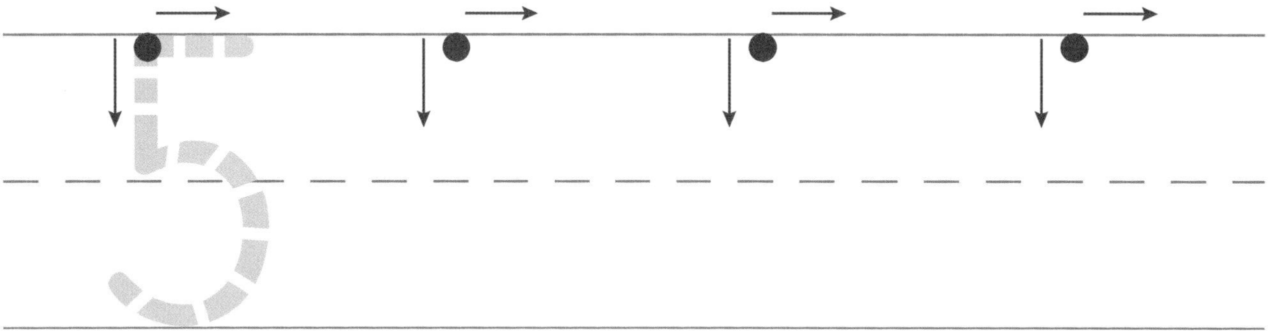

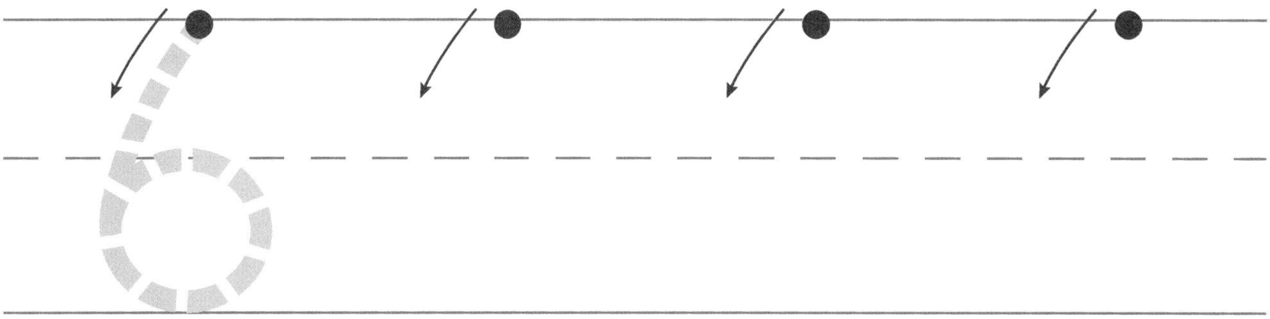

Write 6 and 5.

Chapter 5

Name

 Draw 5   .

 Draw 6  .

 Draw 6  .

Name _____

Color 6 pears on the tree. Color the tree.

Chapter 5

Name _____

Insects have 6 legs. Decorate your insects and give them legs.
Make your own insect on the empty branch.

Chapter 5

Name _____

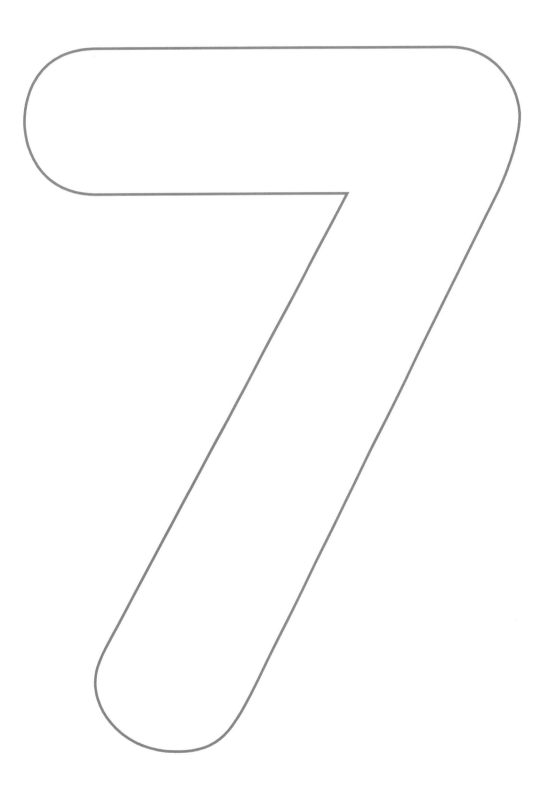

Remove this page from the book. Glue small items inside the seven
(pieces of paper, buttons, sequins, beans, noodles, rice, sand) to make
the seven shape.

Name _____

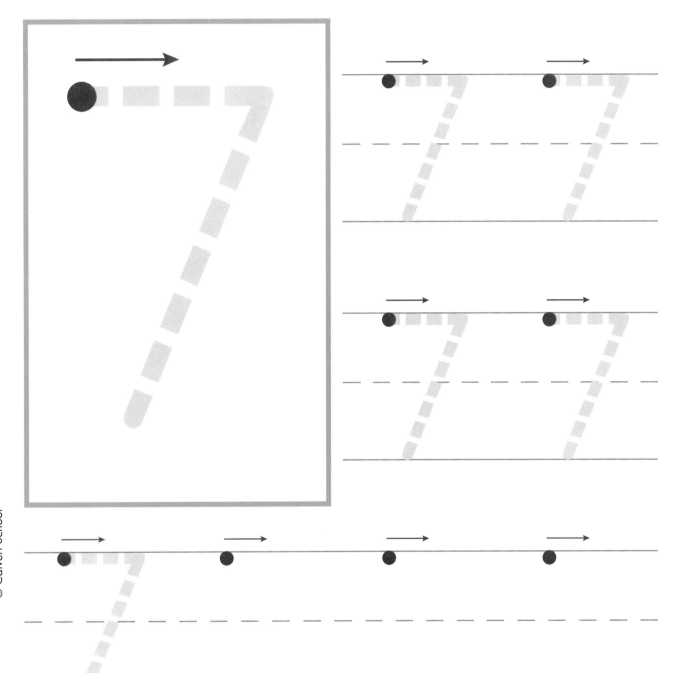

Write 7. Follow the arrow.

Chapter 5

(one hundred nine) **109**

Name _____

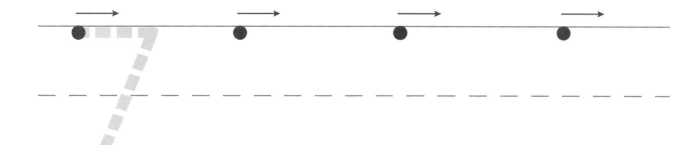

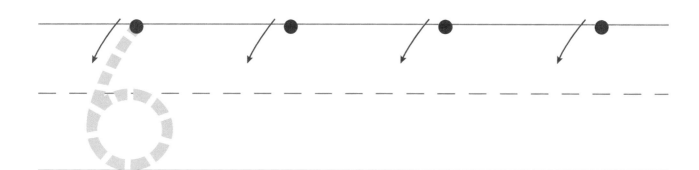

Write 7 and 6. Trace the numbers at the bottom.

Name _____

 Draw 6 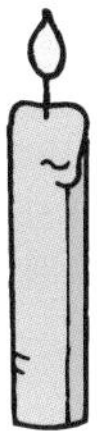  .

Draw 7  .

 Draw 7  .

Name

Color the rabbit. Cut out the carrots. Give the rabbit 7 carrots to eat.

Name

Color each shirt in order.

Name _____

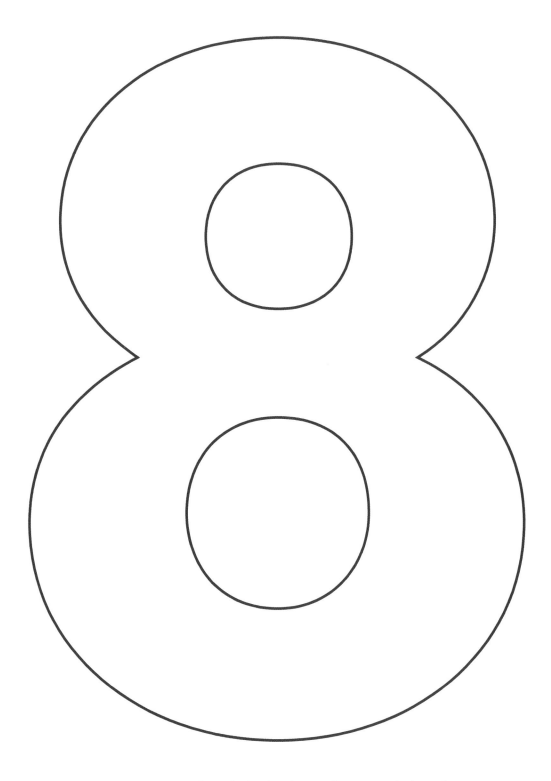

Remove this page from the book. Glue small items inside the eight
(pieces of paper, buttons, sequins, beans, noodles, rice, sand) to make
the eight shape.

Chapter 5

Name _____

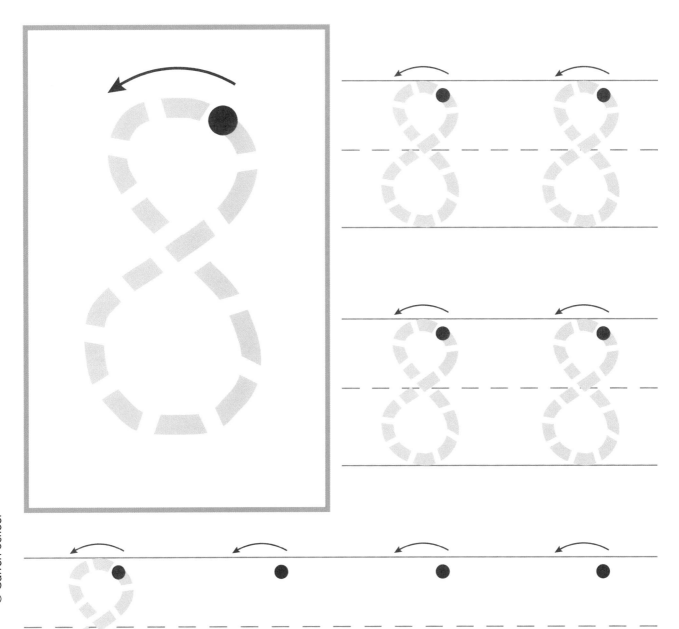

Write 8. Follow the arrow.

Chapter 5 (one hundred nineteen) **119**

Name _____

Draw 8 😊.

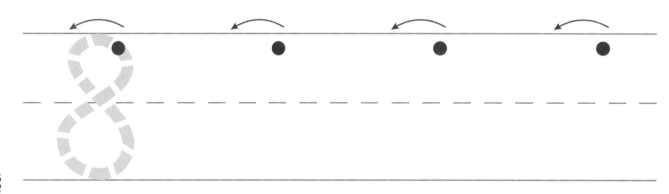

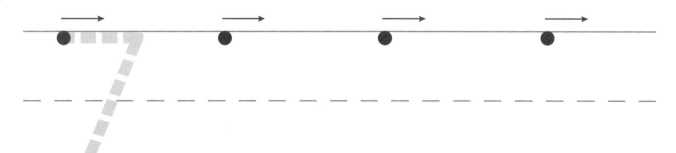

Write 8 and 7.

Chapter 5

Name _____

 Draw 7  .

 Draw 8  .

 Draw 8  .

Name

Color 8 jelly beans in the jar.

Name _____

How many?

🪏									
🧸									
🚗									

_____

_____

_____

Count the items and color one box for each. Write the numbers.

Name _____

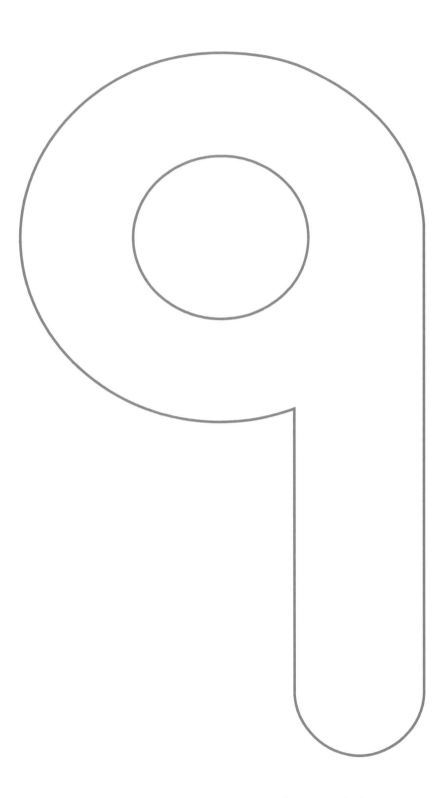

Remove this page from the book. Glue small items inside the nine
(pieces of paper, buttons, sequins, beans, noodles, rice, sand) to make
the nine shape.

Chapter 5 (one hundred twenty-five) **125**

Name _____

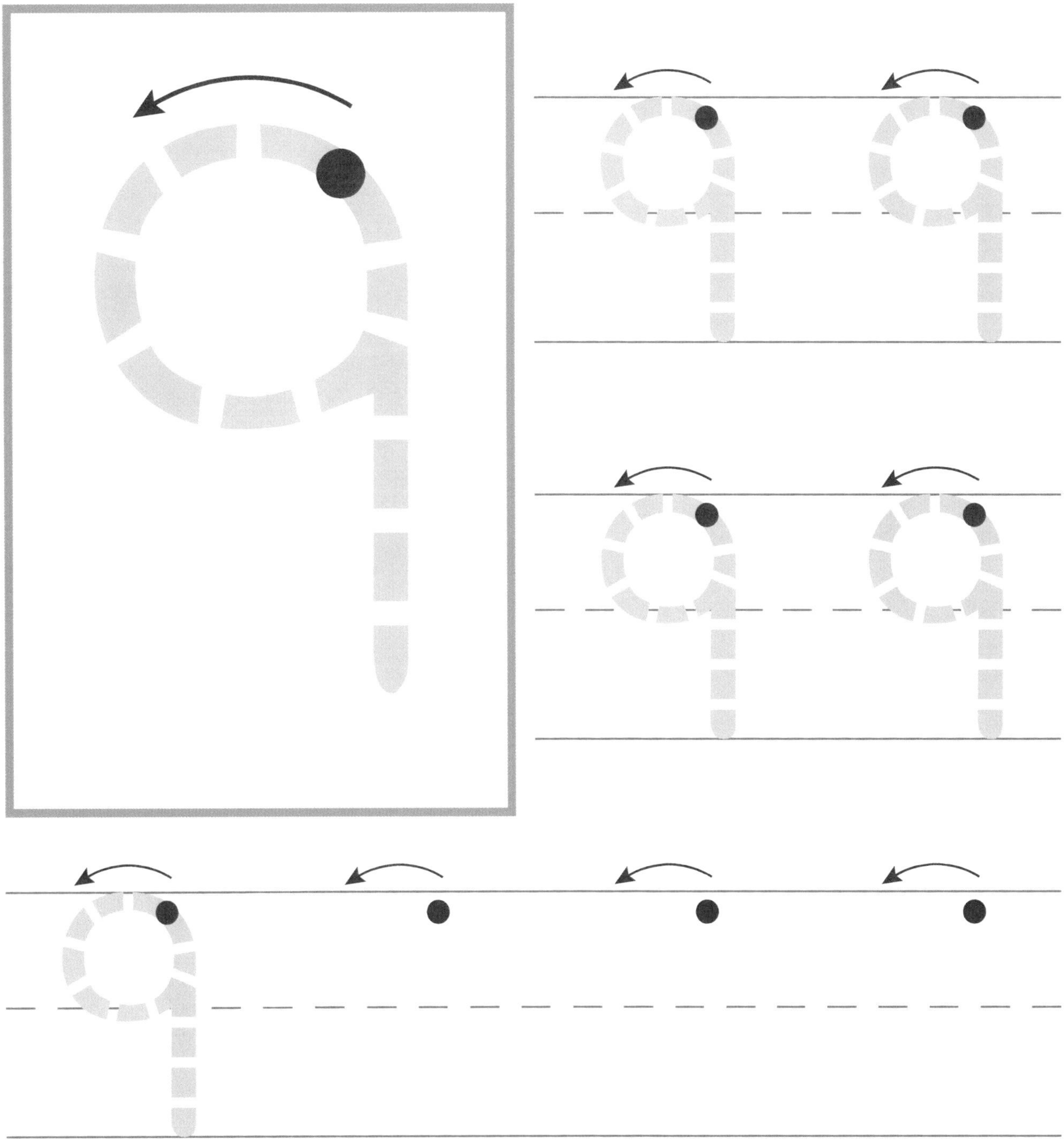

Write 9. Follow the arrow.

Chapter 5

Name _____

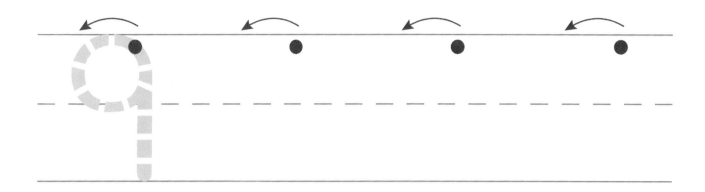

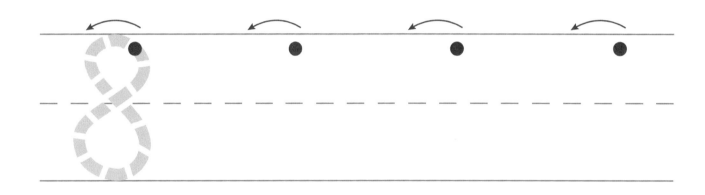

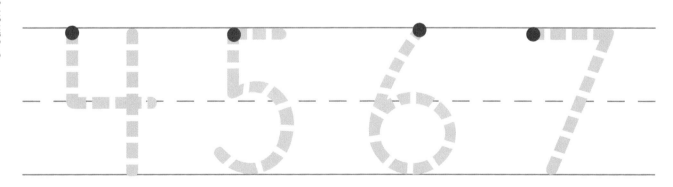

Write 9 and 8. Trace the numbers at the bottom.

Name _____

 Draw 8  .

 Draw 9  .

 Draw 9  .

Name

How many?

_____

_____

_____

_____

Count the pictures and color a box for each.
Write the numbers on the lines.

Chapter 5

© Calvert School

Name _____

Color the picture. Cut out the tulips. Glue 9 tulips in the grass.

Name _____

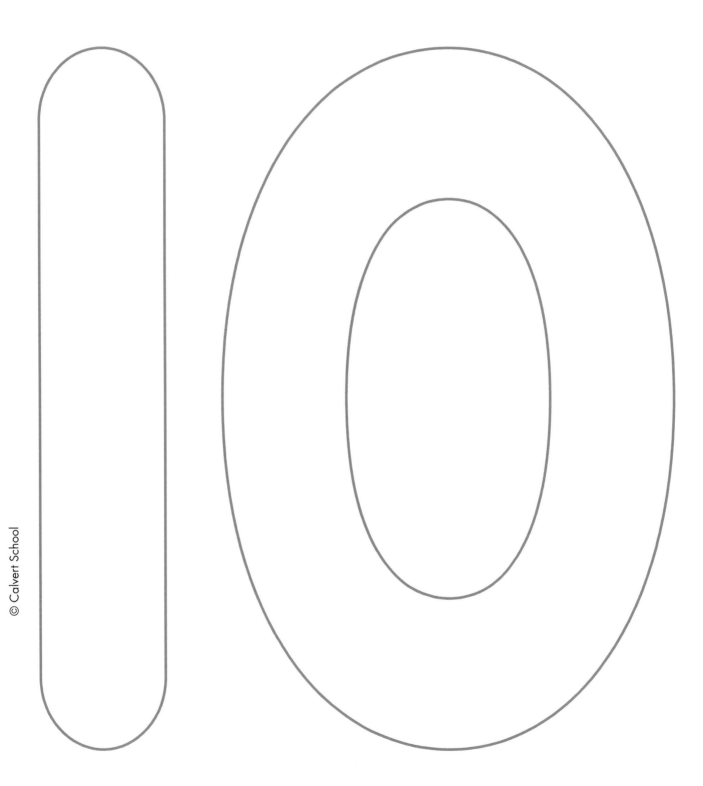

Remove this page from the book. Glue small items inside the ten
(pieces of paper, buttons, sequins, beans, noodles, rice, sand) to make
the ten shape.

Chapter 5

Name

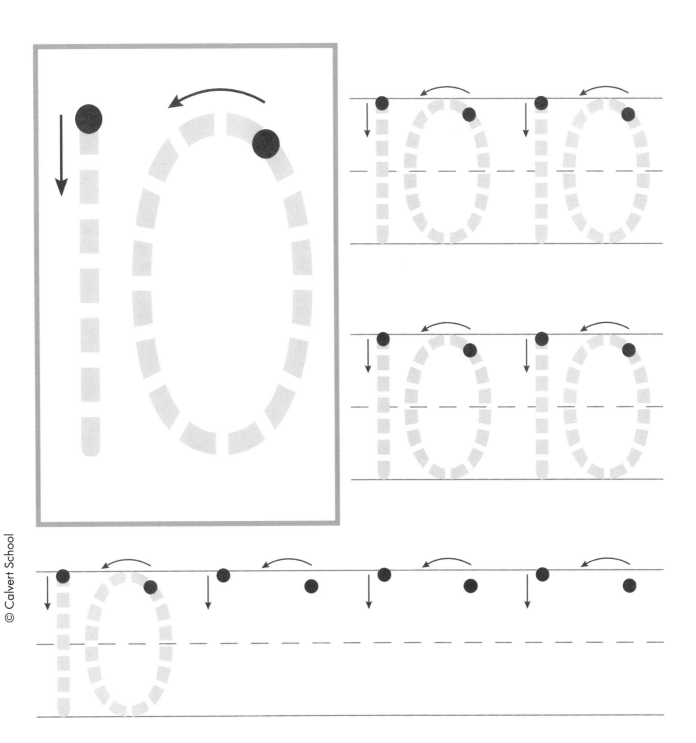

© Calvert School

Write 10. Follow the arrows.

Name

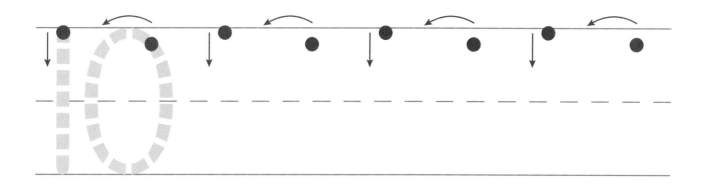

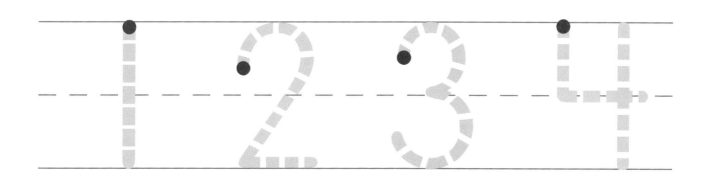

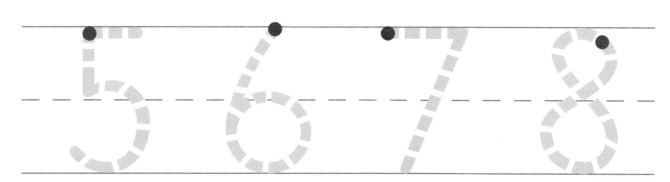

Write 10. Trace the numbers at the bottom.

Name _____

 Draw 9  .

 Draw 10  .

 Draw 10 .

Name

9

10

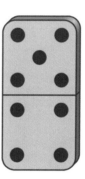

7

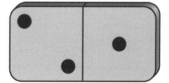

6

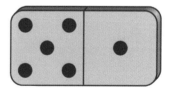

3

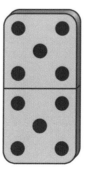

Draw a line to connect the domino with the correct number.

Chapter 5

Name _____

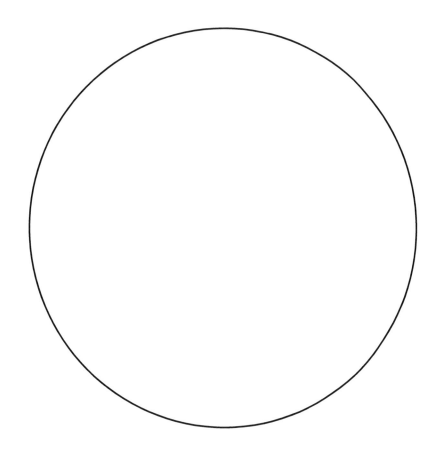

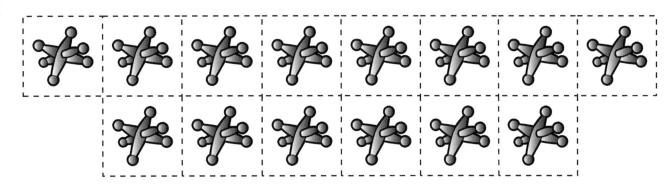

Decorate the ball. Cut out 10 jacks. Arrange the 10 jacks in groups of 2.
Glue the jacks around the ball. How many groups of 2 do you see?

Chapter 5   (one hundred thirty-nine) **139**

Name

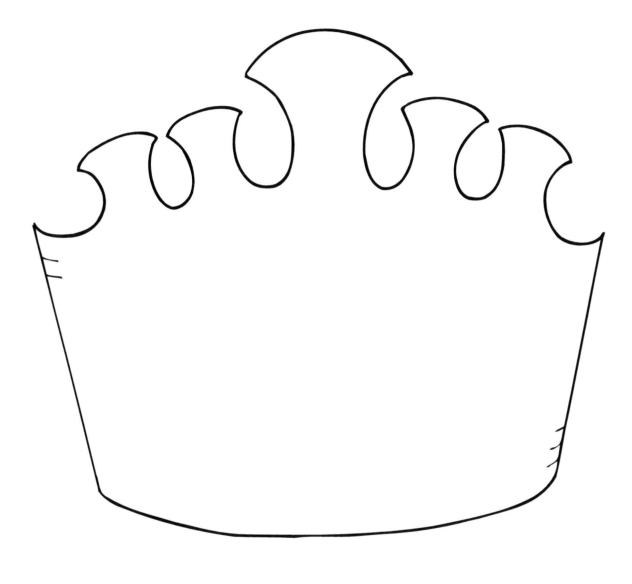

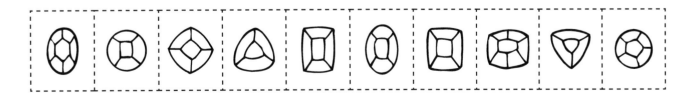

Color the jewels and crown. Cut out 10 jewels. Decorate the regal crown with them.

Name

2
3
4

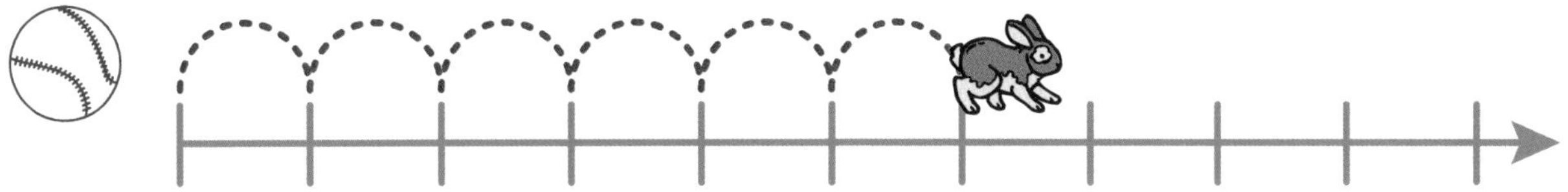

0

4
5
6

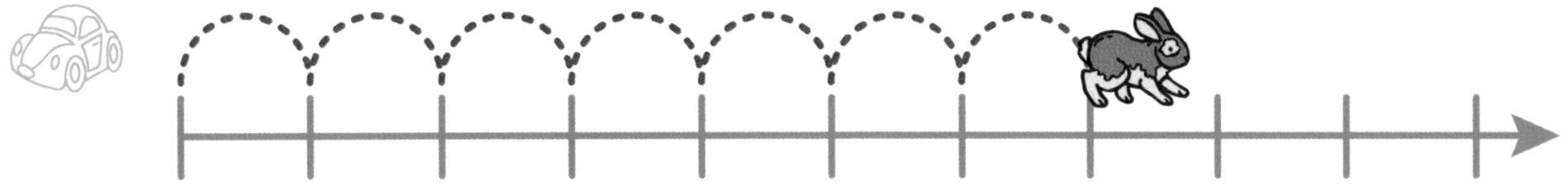

0

7
8
9

0

0
1
2

0

4
5
6

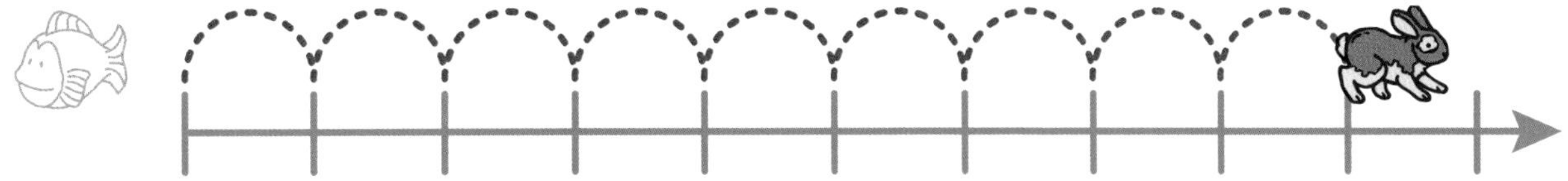

0

8
9
10

0

Circle the numeral that tells how many hops each rabbit has jumped.

Name _____

🍎   3
    4
0   5

⚾   1
    2
0   3

🚗   8
    9
0   10

🐕   3
    4
0   5

🐘   7
    8
0   9

🐟   6
    7
0   8

Circle the numeral that tells how many spaces each frog has hopped.

Chapter 5

My
Number
Book

10

―

9

2

8

3

6

5

Name

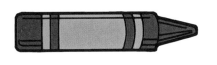

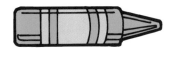

Circle the shortest object.
Mark ✔ on the longest object.

Chapter 6

Name

Mark ✔ on the tallest animal.
Circle the shortest animal.

Chapter 6

Name

Draw an object that is shorter.

Chapter 6

Draw an object that is taller.

Name _____

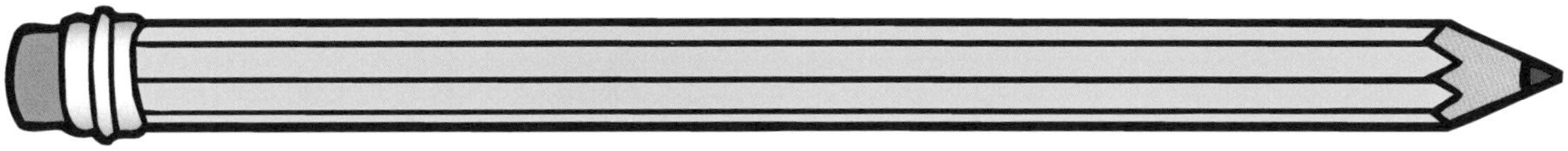

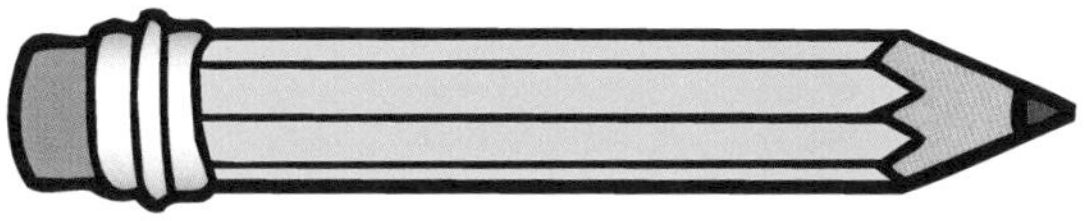

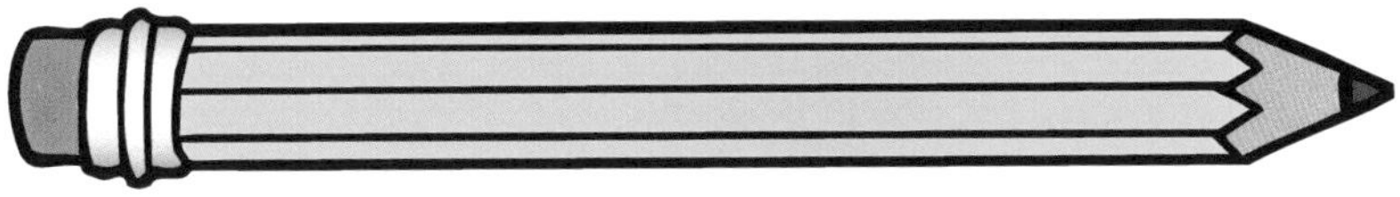

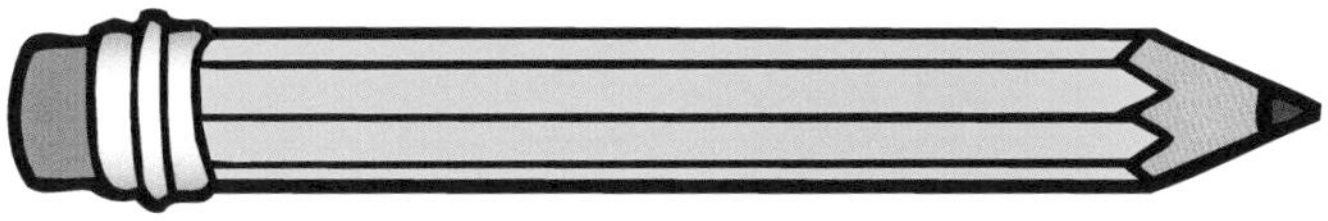

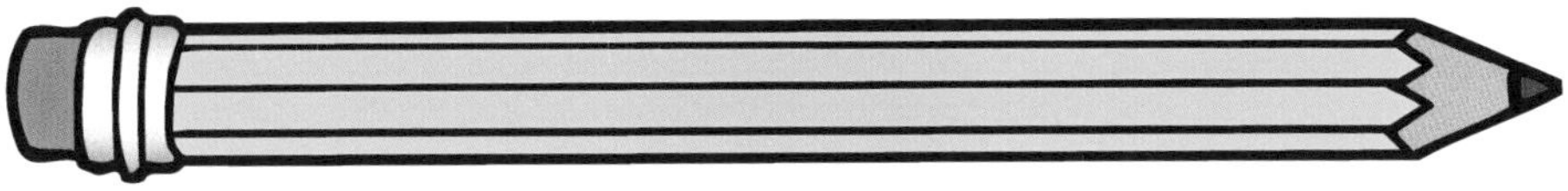

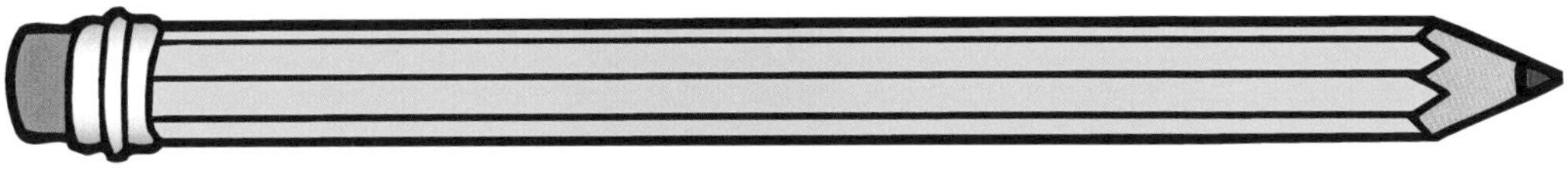

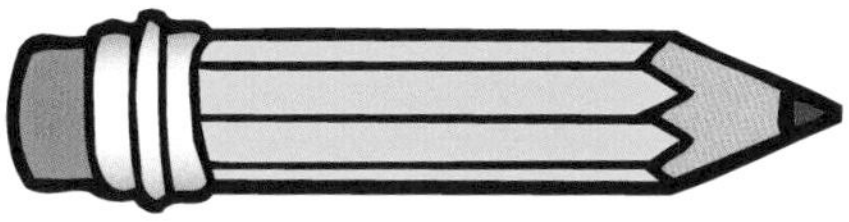

Cut out the pencils and place them in order
from shortest to longest.

Chapter 6

(one hundred fifty-five) **155**

Name _____

Cut out. Glue children on a separate piece of paper in order from
shortest to tallest.

Name _____

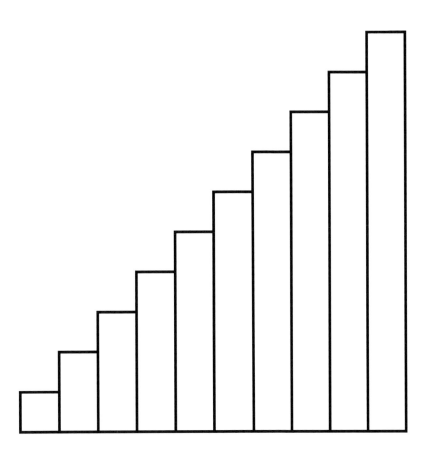

Each rod gets larger by how many rods?
Which rod is shortest? longest?

Name _____

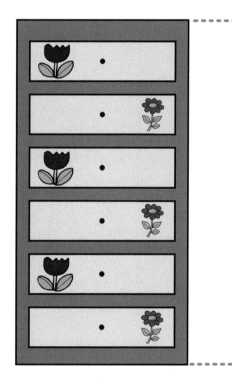

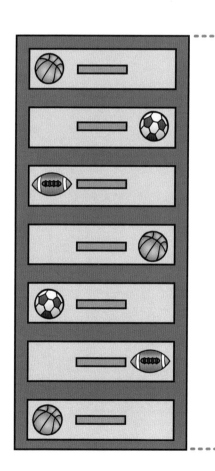

How many cubes? _____   _____

Measure the height of each chest with cubes.
What patterns do you see on the chests?

Name _____

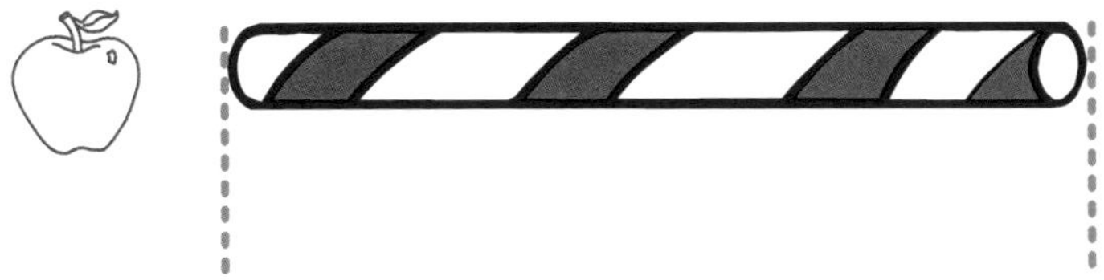

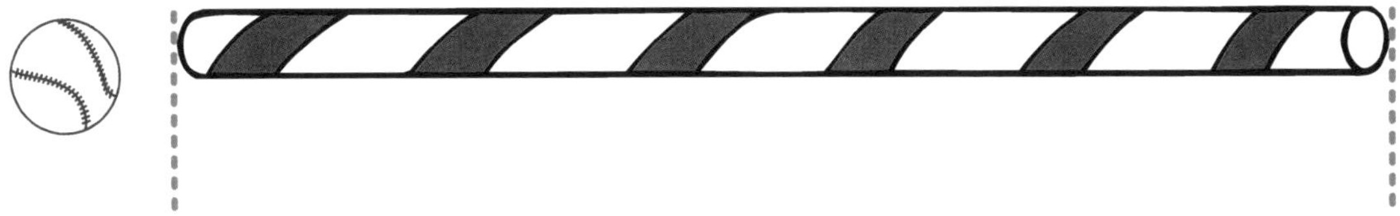

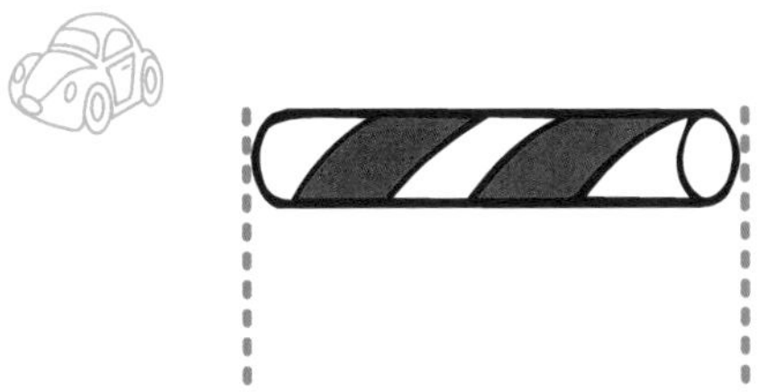

Measure the length of each candy stick with cubes.
Write the number that tells how many cubes.

Name

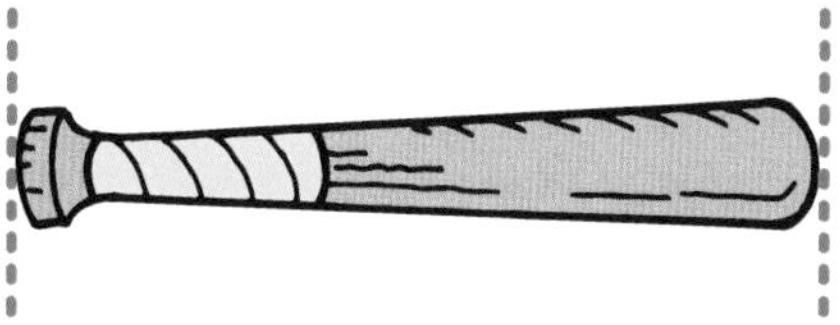

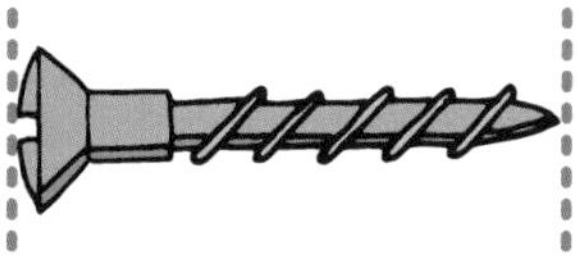

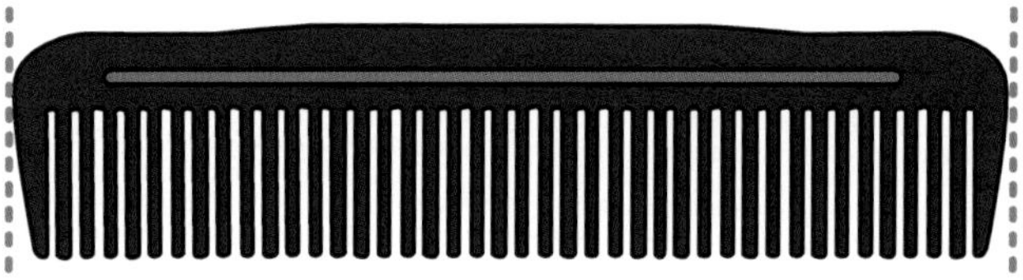

Guess how many cubes. First, write your estimate on the red line. Then use cubes to measure each object. Finally, write how many cubes long on the blue line.

Name _____

Help the children walk through the woods. Trace each path with a crayon. Which path is longer? Why? What could the children see along the way?

Chapter 6

(one hundred sixty-three) **163**

Name _____

Color the containers in order, from holding the most to holding the least.

 blue = 1 (most)   green = 2   yellow = 3 (least)

Chapter 6

Name

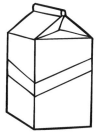

Color the container blue in each row that holds the most.
Color the container orange in each row that holds the least.

Name

Circle the picture in each row that is the heaviest.

Name

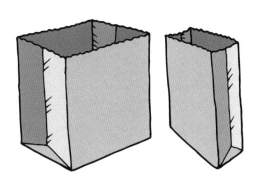

Circle the object that holds more. Draw a line
under the object that holds less.

Chapter 6

Name

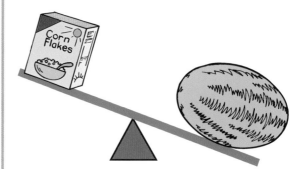

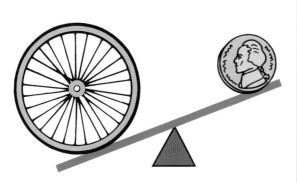

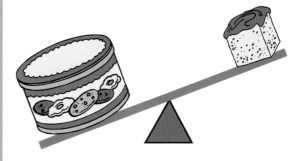

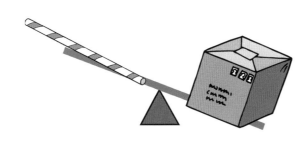

Circle the object that weighs less.

Chapter 6

Name _____

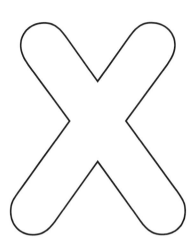

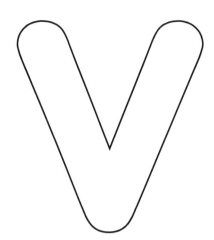

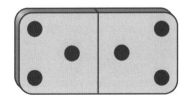

Chapter 6

Draw a line through the middle of each picture to make
2 equal parts.

(one hundred sixty-nine) **169**

Name

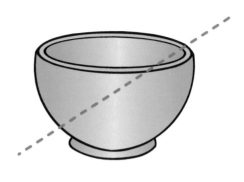

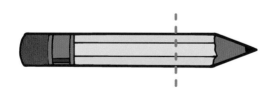

Circle the picture that shows 2 equal parts.

Chapter 6

Name _____

 If two parts of a shape match exactly when folded, the shape is symmetrical.

 Make a symmetrical design to match the one on the right.

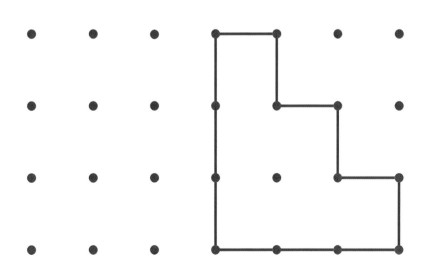

Name

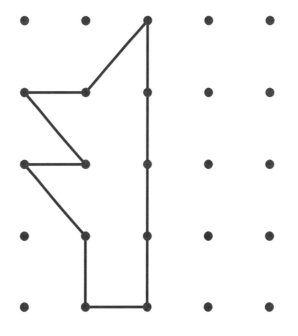

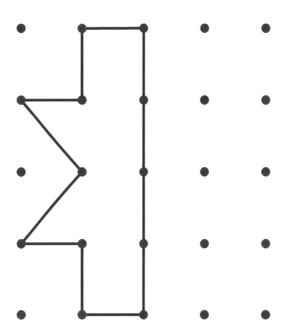

Draw more symmetrical shapes by connecting dots on the right to
match the shape on the left.

Name

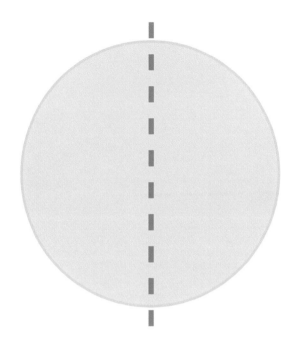

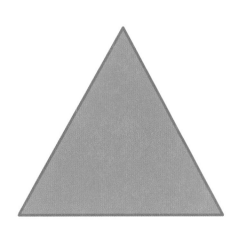

When shapes are divided into 2 equal parts, each part is
called one half or $\frac{1}{2}$. Draw a line to show 2 equal parts.

Name _____

A Sweet Graph

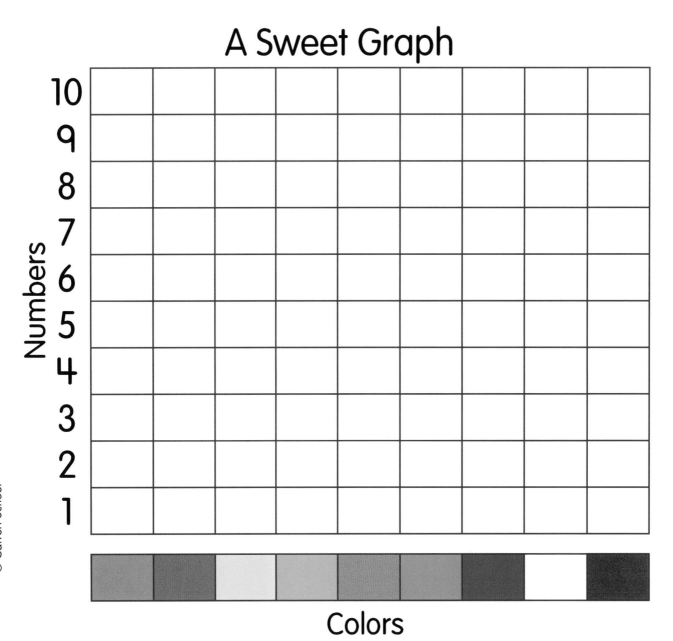

Graph by color the marshmallow shapes in one bowl of cereal or the
candies from a small bag. Then eat them!

Chapter 6

Name

Draw a line from the season to the matching objects.

Chapter 6

(one hundred seventy-five) **175**

Name

Cut out the apples. Use them to tell number stories.

Name _____

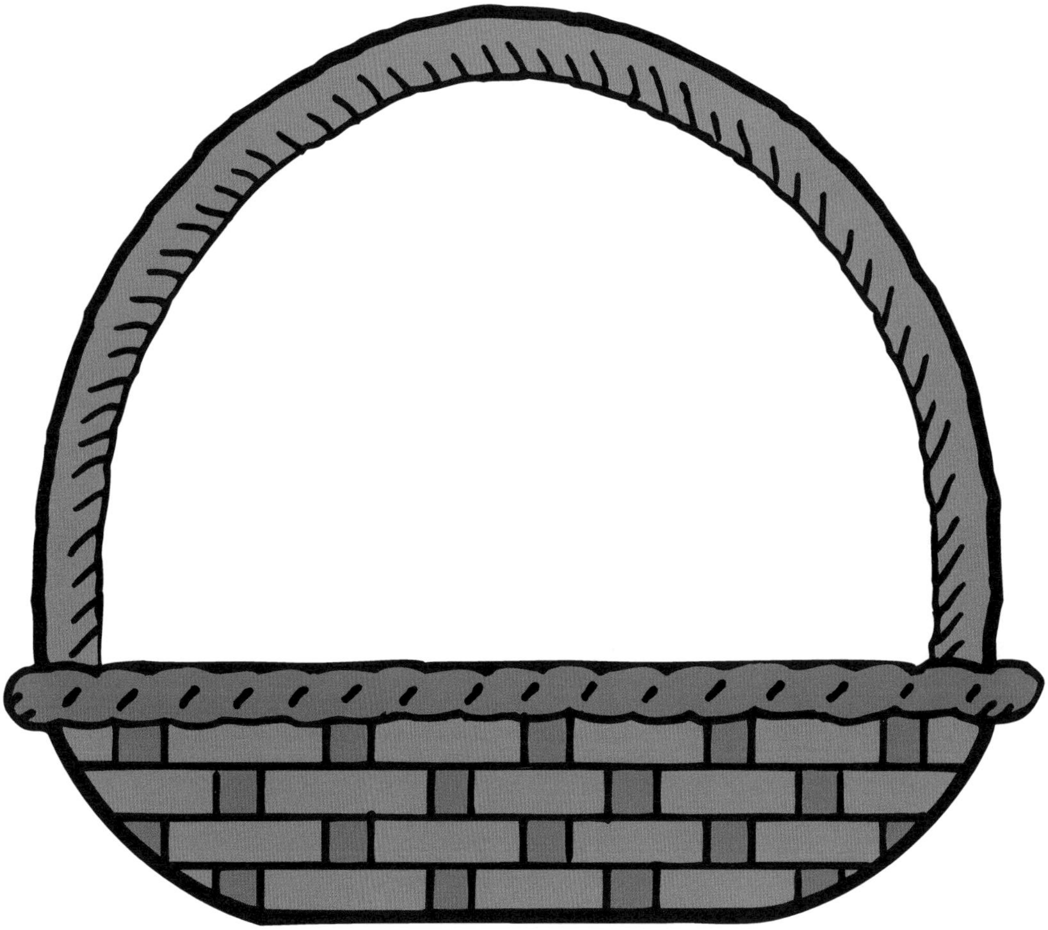

Cut out the apples. Use them to tell number stories.

178 (one hundred seventy-eight)

Chapter 7

Name _____

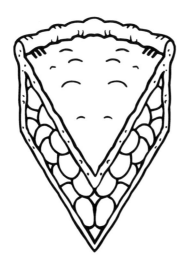

Draw a line through each object to make two parts of the same size. Color one half.

Chapter 7

Name _____

_____　　　　_____　　　　_____

_ _ _ _ _ _ _　**+**　_ _ _ _ _ _ _　**=**　_ _ _ _ _ _ _

_____　　　　_____　　　　_____

_____　　　　_____　　　　_____

_ _ _ _ _ _ _　**+**　_ _ _ _ _ _ _　**=**　_ _ _ _ _ _ _

_____　　　　_____　　　　_____

_____　　　　_____　　　　_____

_ _ _ _ _ _ _　**+**　_ _ _ _ _ _ _　**=**　_ _ _ _ _ _ _

_____　　　　_____　　　　_____

Tell the number story. Write the numbers that tell the number story.

Name _____

Draw one more and write the number of objects.

_____

- - - - - - - - - - - - -

_____

_____

- - - - - - - - - - - - -

_____

Draw one more and write the number.

_____

- - - - - - - - - - - - -

_____

_____

- - - - - - - - - - - - -

_____

Write the number before and after.

- - - - - - - 5 - - - - - - - - - - 1 - - - - - - -

Chapter 7 (one hundred eighty-one) **181**

Name

_____   _____   _____

- - - - - - -   **+**   - - - - - - -   **=**   - - - - - - -

_____   _____   _____

_____   _____   _____

- - - - - - -   **+**   - - - - - - -   **=**   - - - - - - -

_____   _____   _____

_____   _____   _____

- - - - - - -   **+**   - - - - - - -   **=**   - - - - - - -

_____   _____   _____

Tell the number story. Write the number sentence.

Name

| $4 + 2 = 6$ | $5 + 1 = 6$ | $2 + 4 = 6$ | $2 + 3 = 5$ |

| $2 + 3 = 5$ | $3 + 2 = 5$ | $1 + 4 = 5$ | $4 + 1 = 5$ |

| $3 + 3 = 6$ | $2 + 4 = 6$ | $1 + 4 = 5$ | $4 + 1 = 5$ |

Put an X on the box that tells the number story.

Chapter 7

Name _____

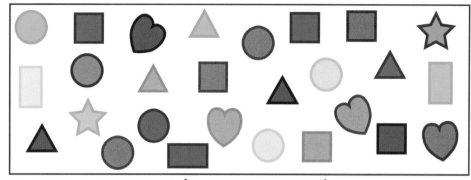

Shape Graph

Number of Shapes

10
9
8
7
6
5
4
3
2
1

♡	□	○	▭	◁	☆

Shapes

Count the shapes. Graph the number of each shape.

Name

_____   _____   _____

- - - - - -   +   - - - - - -   =   - - - - - -
_____   _____   _____

_____   _____   _____

- - - - - -   +   - - - - - -   =   - - - - - -
_____   _____   _____

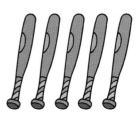

_____   _____   _____

- - - - - -   +   - - - - - -   =   - - - - - -
_____   _____   _____

Tell the number story. Write the number sentence.

Chapter 7

Name _____

 △ △ △ △ △ △ △ △ △ △

_____

$3 + 2 =$

 △ △ △ △ △ △ △ △ △ △

_____

$4 + 1 =$

 △ △ △ △ △ △ △ △ △ △

_____

$1 + 3 =$

 △ △ △ △ △ △ △ △ △ △

_____

$2 + 1 =$

Color the △ to show the number sentence. Write the answer.

Name _____

4 + 4 = 8 ●

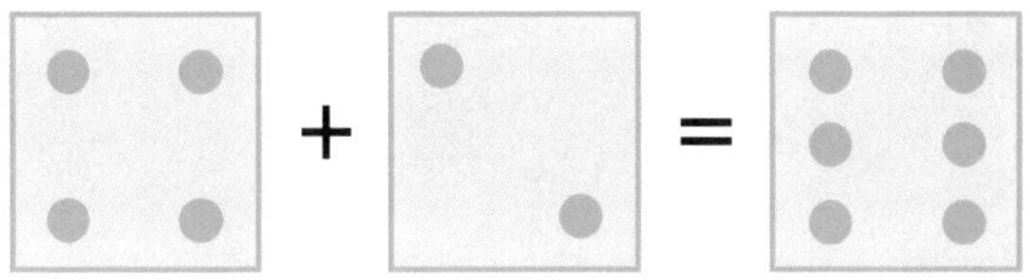

3 + 5 = 8 ●

4 + 2 = 6 ●

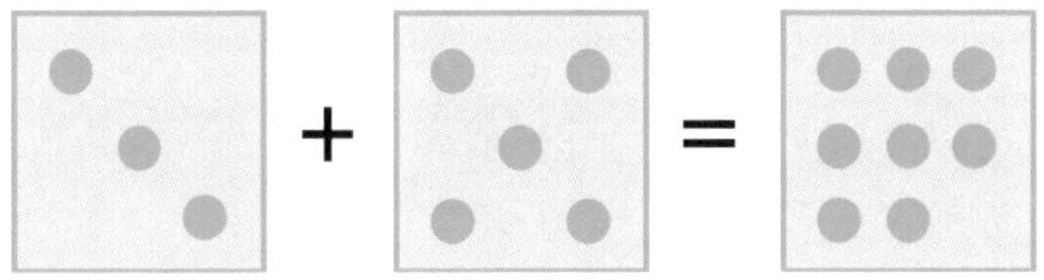

5 + 1 = 6 ●

● ▢ + ▢ = ▢

● ▢ + ▢ = ▢

● ▢ + ▢ = ▢

● ▢ + ▢ = ▢

Draw a line to match the number sentence with the picture that shows
the number sentence.

Name

2 + 4 = 6 ⬤

⬤ [die: 5] + [die: 1] = [die: 6]

3 + 3 = 6 ⬤

⬤ [die: 3] + [die: 2] = [die: 5]

3 + 2 = 5 ⬤

⬤ [die: 3] + [die: 3] = [die: 6]

5 + 1 = 6 ⬤

⬤ [die: 2] + [die: 4] = [die: 6]

Draw a line to match the number sentence to the picture.

Chapter 7

Name _____

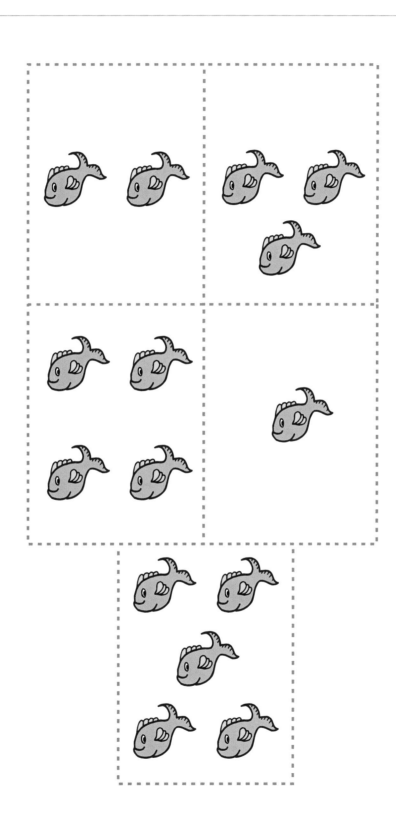

Cut out the number cards. Make a number story.

Name _____

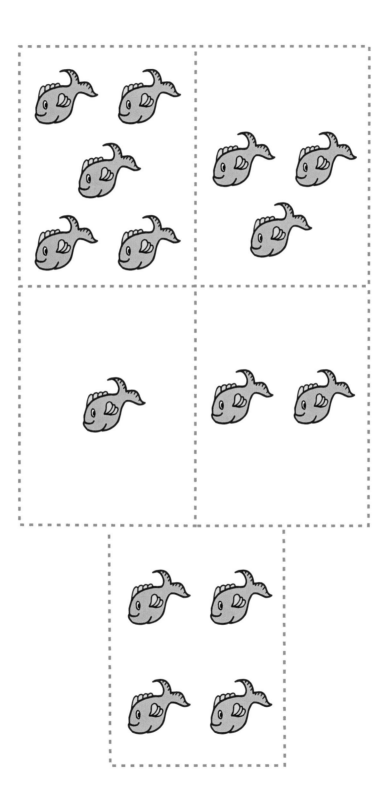

Cut out the number cards. Make a number story.

Name _____

 10 | 13   5   3   11

 3 | 0   9   7   1

 8 | 4   6   10   13

 6 | 9   8   3   2

 5 | 4   9   0   6

 9 | 12   10   4   7

Put an X on all of the numbers that come before the first number.

Name

3	2			0

5	9	4	6	1

8	10	2	7	9

2	8	0	1	4

6	2	3	9	6

9	11	12	5	8

Put an X on all of the numbers that come after the first number.

Name _____

_____ — _____ = _____

_____ — _____ = _____

_____ — _____ = _____

Tell the number story. Write the numerals that tell the number story.

Name

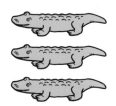  

_____   _____   _____

- - - - - - -   ▬   - - - - - - -   ═   - - - - - - -

_____   _____   _____

_____   _____   _____

- - - - - - -   ▬   - - - - - - -   ═   - - - - - - -

_____   _____   _____

_____   _____   _____

- - - - - - -   ▬   - - - - - - -   ═   - - - - - - -

_____   _____   _____

Tell the number story. Write the numerals that tell the number story.

Draw a line through one object in each box.
Write the number of the remaining objects.

Chapter 8

(one hundred ninety-seven) **197**

Name _____

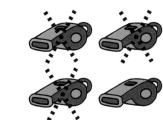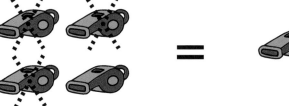

| $4 - 2 = 2$ | $4 - 3 = 1$ | $4 - 1 = 3$ | $4 - 0 = 4$ |

| $6 - 2 = 4$ | $6 - 4 = 2$ | $6 - 1 = 5$ | $6 - 3 = 3$ |

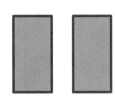

| $4 - 3 = 1$ | $4 - 1 = 3$ | $4 - 2 = 2$ | $4 - 0 = 4$ |

Tell the number story. Circle the correct number sentence.

Name _____

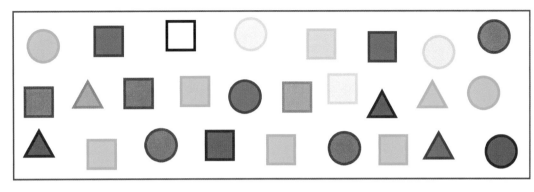

Color Graph

10						
9						
8						
7						
6						
5						
4						
3						
2						
1						

Number of Shapes

Blue	Red	Orange	Yellow	Green	Purple	White

Colors

Graph the shapes by color.

© Calvert School

Name _____

$$6 - 2 =$$

_____

- - - - - - -

_____

$$5 - 5 =$$

_____

- - - - - - -

_____

$$3 - 2 =$$

_____

- - - - - - -

_____

$$4 - 3 =$$

_____

- - - - - - -

_____

Put an X on each to show the number taken away.
Write the number of hearts left.

Chapter 8

Name _____

$6 - 1 = 5$ ●   ●

$5 - 3 = 2$ ●   ●

$4 - 1 = 3$ ●   ●

$6 - 6 = 0$ ●   ●

Match the number sentence to the number story.

Chapter 8   (two hundred one) **201**

Name _____

8   9 | 2

|2 |3   2   3

7   2   5   6

4   3 || |2

8   5   6   9

In each box draw a line under the number that is more.

Name _____

🍎	5 _____	7 _____	
⚾	9 _____	1 _____	
🚗	2 _____	6 _____	
🐕	8 _____	3 _____	
🐘	4 _____	0 _____	

Write the number that comes right after the
number on the left.

Chapter 8

Name _____

Begin Here.

Make the first two faces red.
Make the next three faces yellow.
Make the next five faces purple.
Make the next one face blue.
Make the next six faces orange.
Make the next four faces brown.
How many faces are not colored?

© Calvert School

Name

 Trace the 11. Write the number 11.

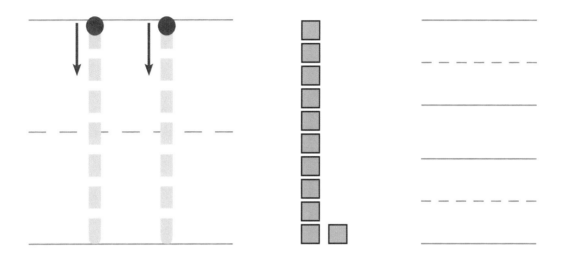

Color 11 balloons.

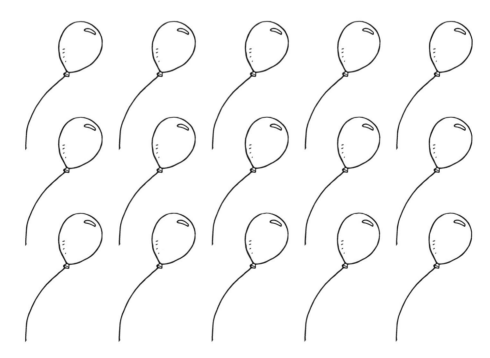

Name

There are eleven children on the soccer team.
Number their shirts from 1 to 11.

Chapter 9

Name _____

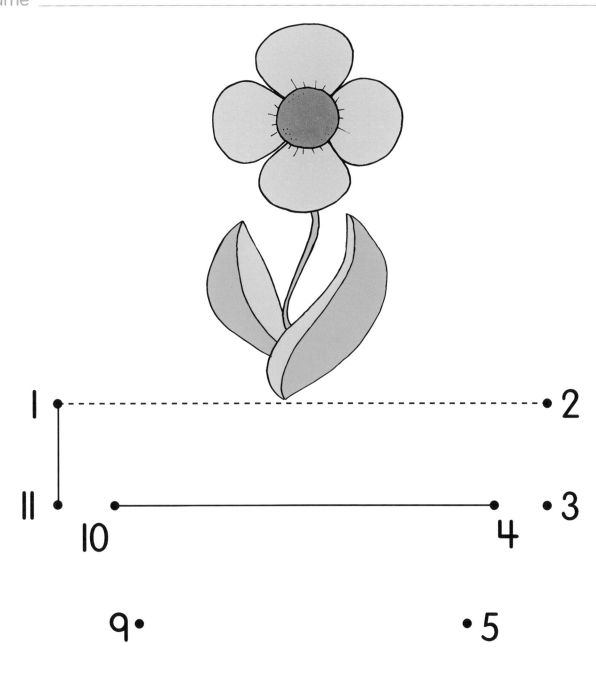

Use the numbers to connect the dots.

Name _____

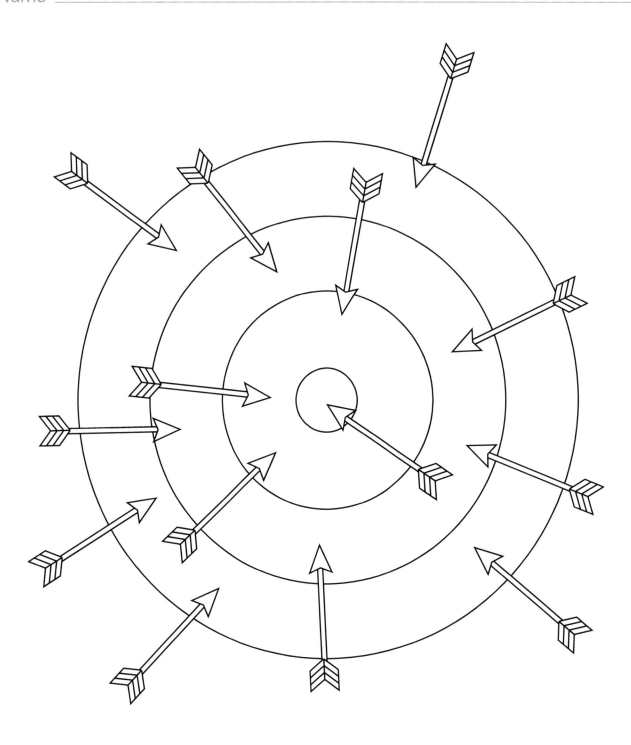

Color 11 arrows on the target.

Chapter 9

Name

 Trace the 12. Write the number 12.

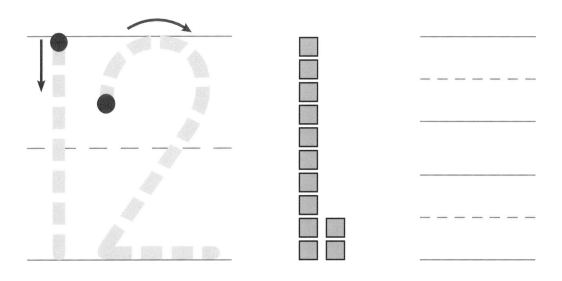

Color 12 crayons.

Name

Color the bird feeder. Cut out the birds. Glue 12 birds eating seeds.

Name

 Trace the 13. Write the number 13.

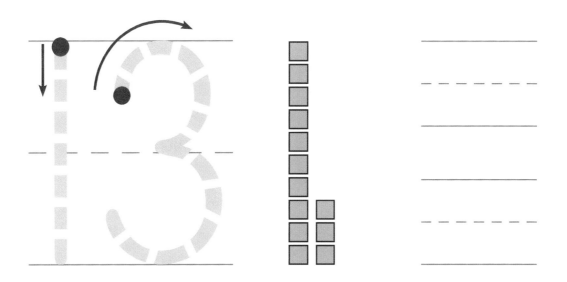

🏐 Color 13 triangles.

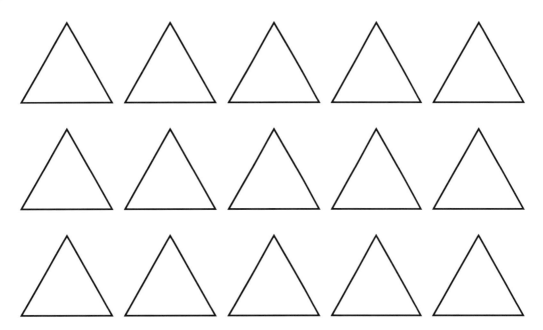

Name _____

Color 13 spiders on the web.

Chapter 9

Name

 Trace the 14. Write the number 14.

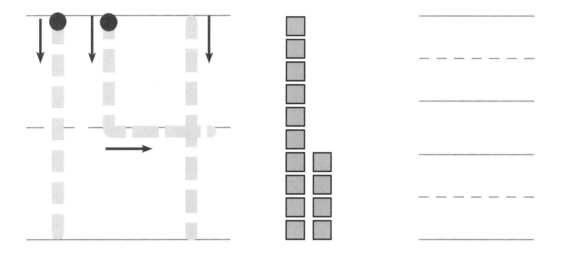

 Color 14 leaves.

Name _____

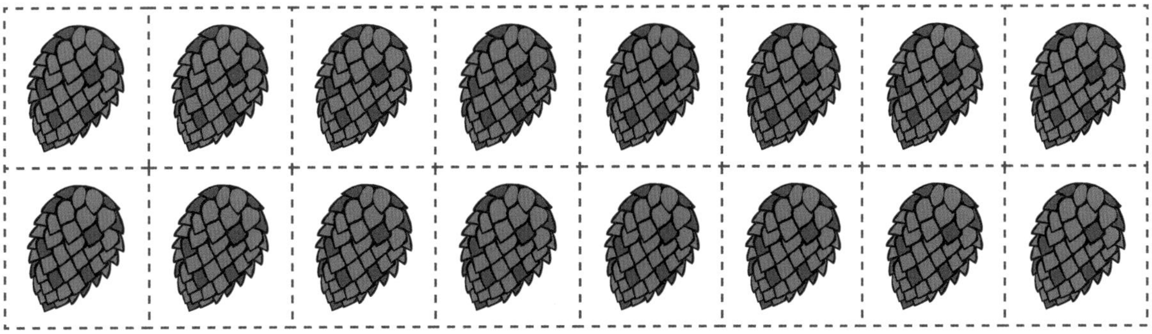

Color the pine tree. Cut out the pinecones.
Glue 14 pine cones on the tree.

Name

 Trace the 15. Write the number 15.

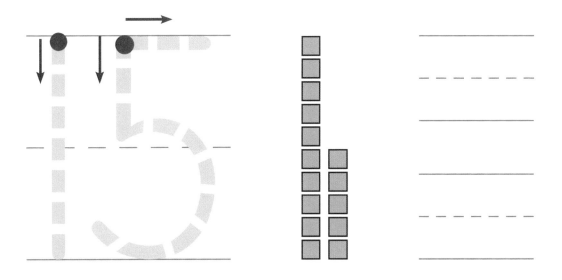

Color 15 bananas.

Name

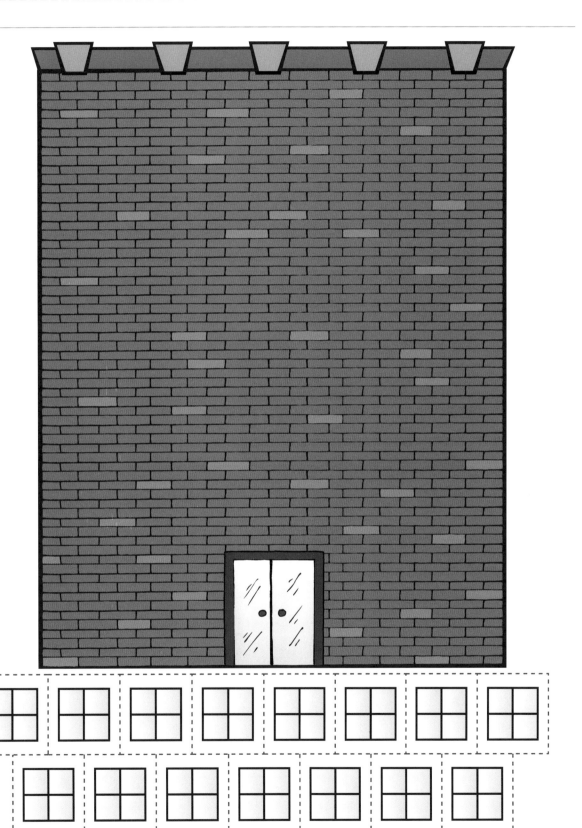

Cut out and then glue 15 windows on the apartment building.

Name

Candy Graph

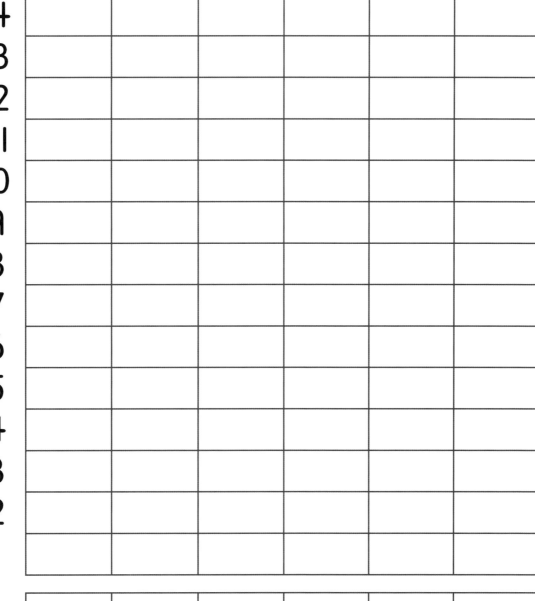

Numbers

15
14
13
12
11
10
9
8
7
6
5
4
3
2
1

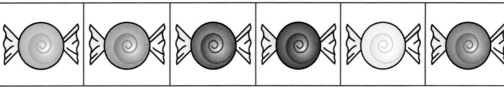

Colors

Sort cereal pieces or candy by color. Count each color.
Show each number on the graph. (two hundred twenty-three)

Name

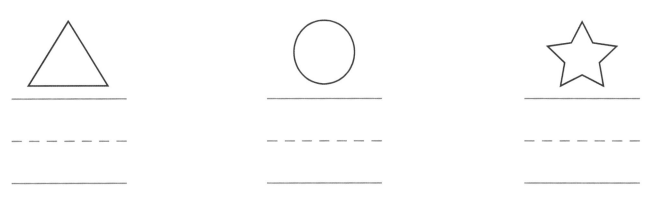

Count the shapes. Write the number.

Chapter 9

Name

 Trace the 16. Write the number 16.

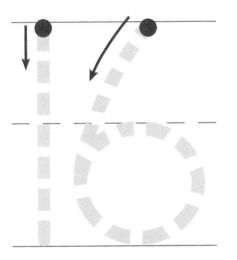 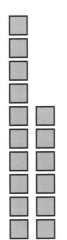 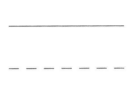

 Color 16 ice cream cones.

Name _____

Cut out 16 fish from p. 227, then glue them in the fishbowl.

Chapter 9

Name _____

Name

 Trace the 17. Write the number 17.

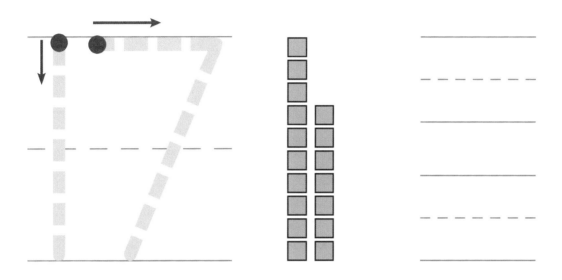

 Color 17 mittens.

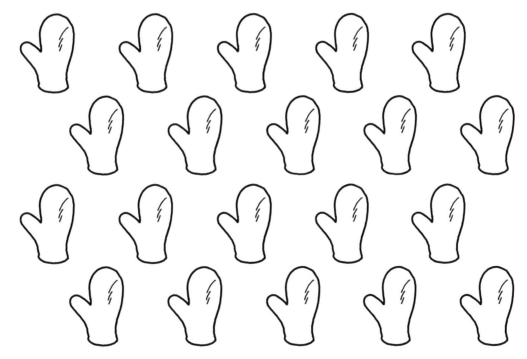

Name

17 people are coming to the birthday party. Draw more plates to make one for each person.

Chapter 9

Name

Trace the 18. Write the number 18.

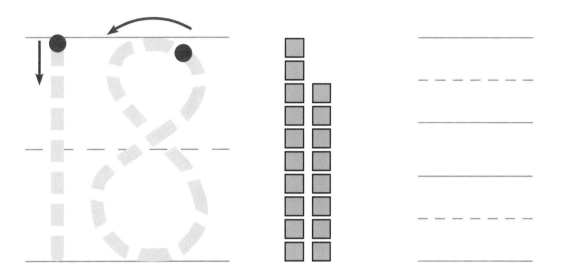

Color 18 apples.

Name

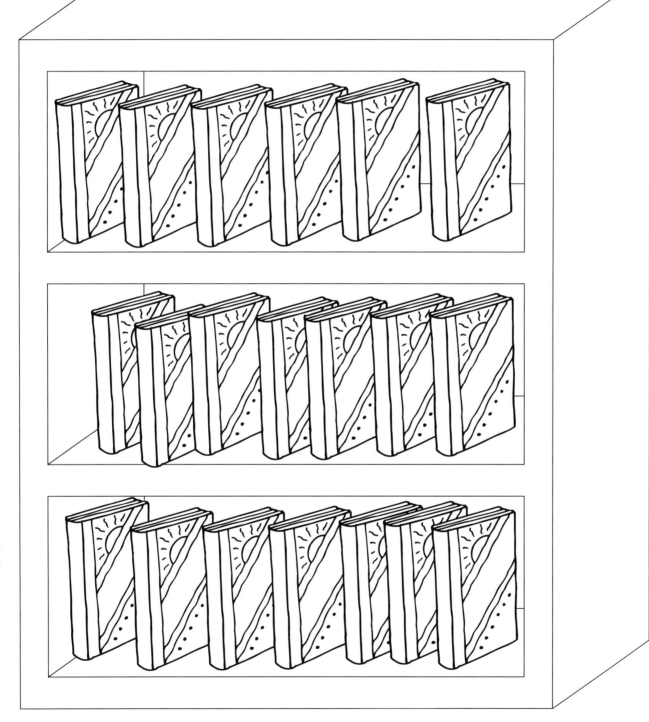

Color 18 books on the bookcase.

Name

Trace the 19. Write the number 19.

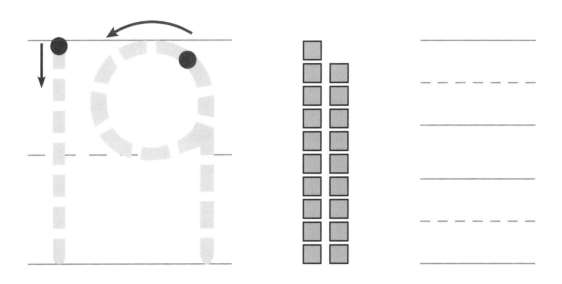

Color 19 carrots.

Name _____

Color 19 blueberries on the bush.

Name

 Trace the 20. Write the number 20.

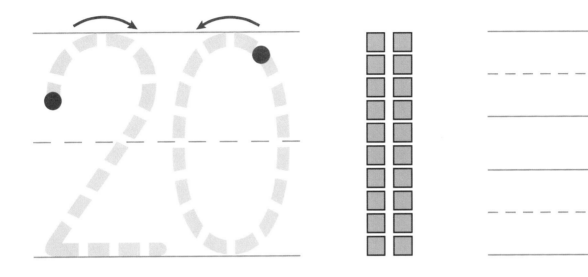

 Color 20 pencils.

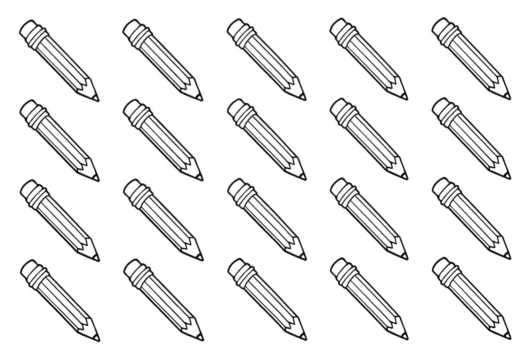

© Calvert School

Name _____

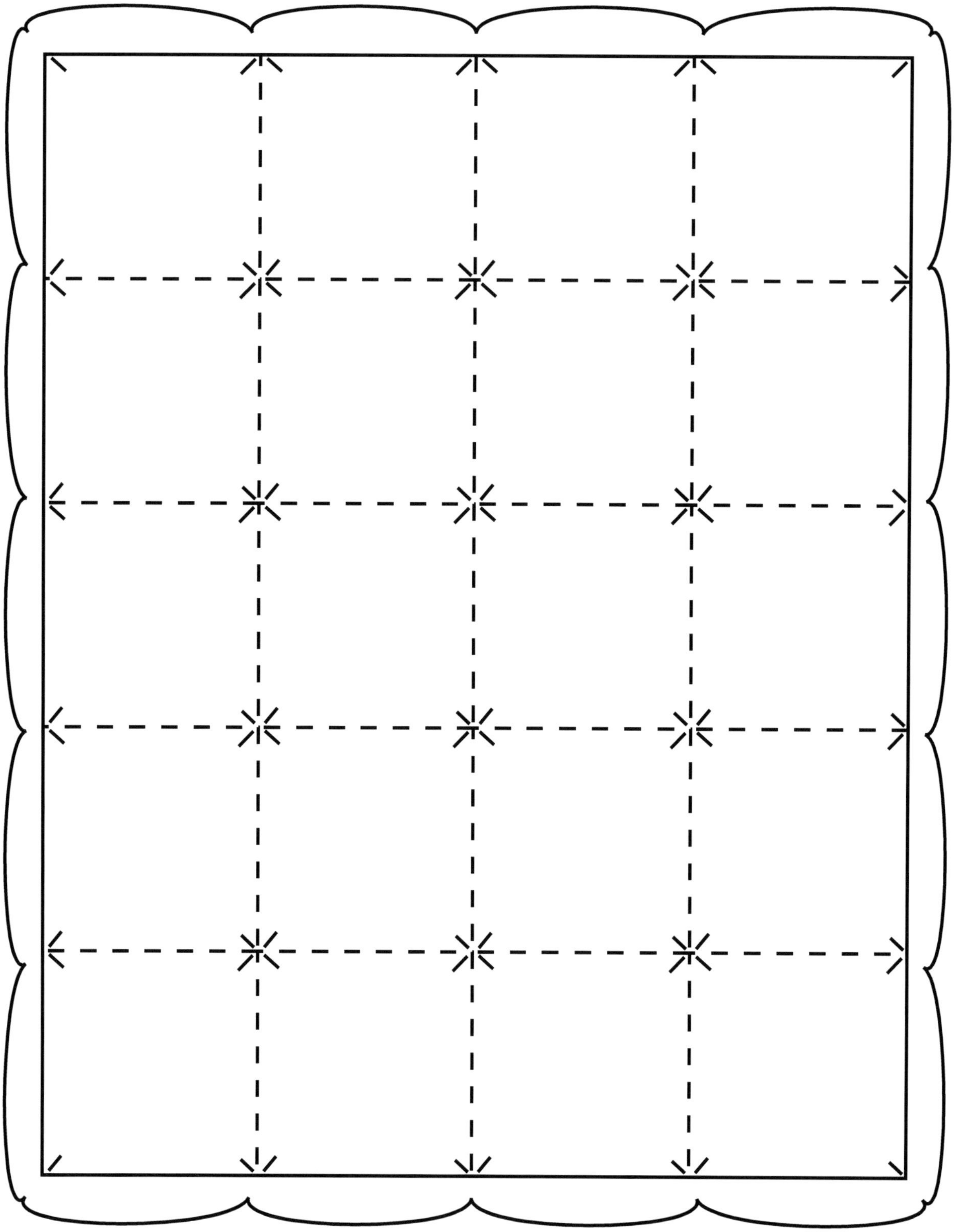

The quilt has 20 squares. Use three colors to make a pattern by coloring the squares.

Chapter 9

Name

1 one	
2 two	
3 three	
4 four	
5 five	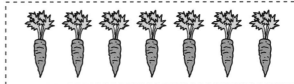
6 six	
7 seven	
8 eight	

Cut out the numbers and vegetable pictures along the dotted lines.
Count the vegetables. Glue the vegetables next to the matching
number on a separate sheet of paper.

Chapter 9   (two hundred thirty-seven) **237**

Name

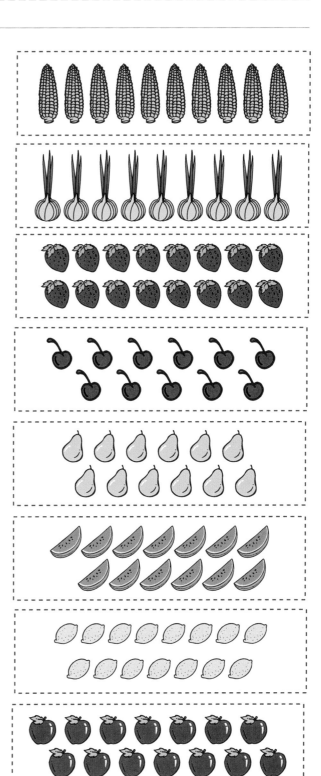

9 nine

10 ten

11 eleven

12 twelve

13 thirteen

14 fourteen

15 fifteen

16 sixteen

Cut out the numbers and fruit pictures along the dotted lines.
Count the fruit. Glue the fruit next to the matching number on a
separate sheet of paper.

Name

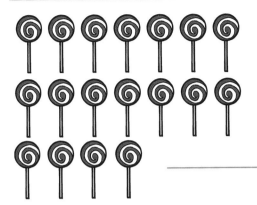
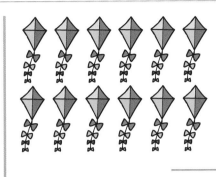

_____

- - - - - - - - - - -

_____

_____

- - - - - - - - - - -

_____

_____

- - - - - - - - - - -

_____

Count the objects and write how many.

Chapter 9

Name _____

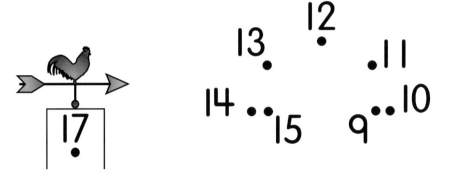

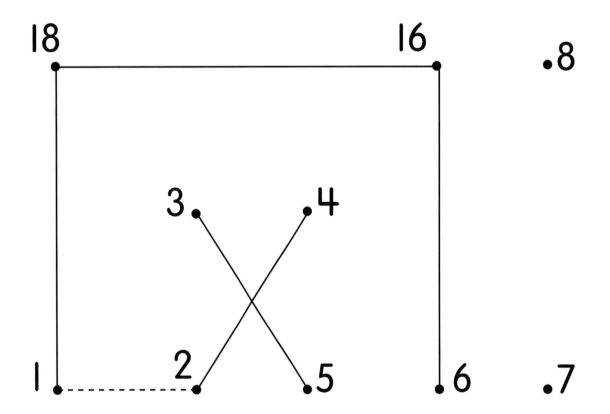

Start at 1. Connect the dots in order.

© Calvert School

Name _____

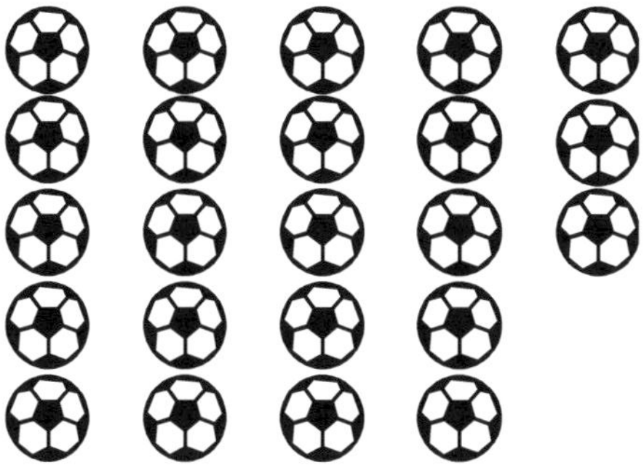

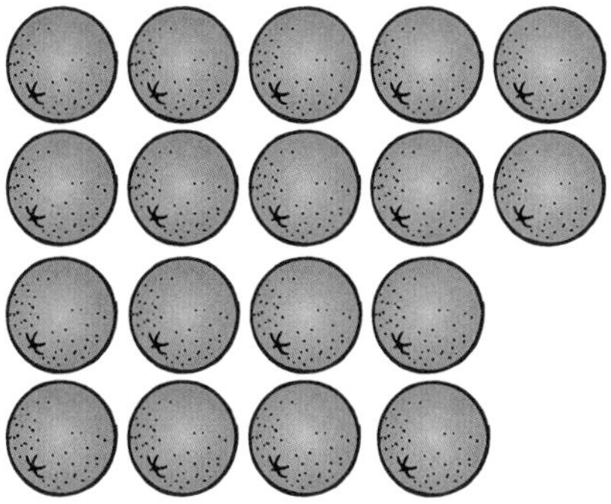

Count the objects. Then write the number.

Chapter 9

(two hundred forty-three) **243**

Name _____

Count the objects. Then write the number.

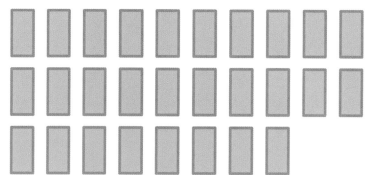

Name _____

1

28

8   11

2   9   10
27   7

4   5   6

3

26

25

24   21   20   17

16   15   14   13

23 22   19 18

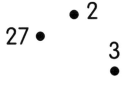
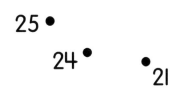
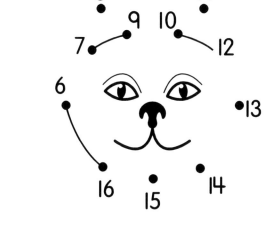

Start at 1 and connect the dots.

Name

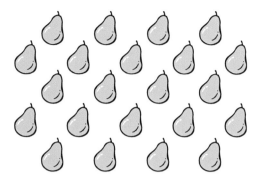 24

 17

 30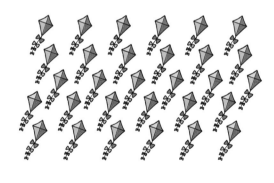

 10

Without counting, circle the picture that shows about how many.

Chapter 9

Name _____

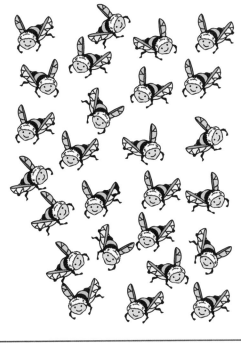

5	10	30

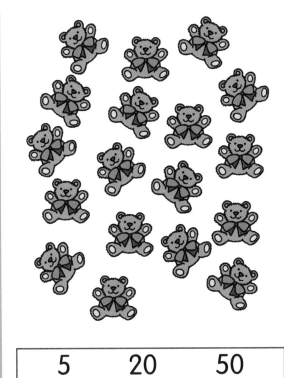

5	20	50

5	10	50

10	30	100

Without counting, circle the number
that shows about how many.

Name

_____   _____   _____

- - - - -   - - - - -   - - - - -

_____   _____   _____

- - - - -   - - - - -   - - - - -

_____   _____   _____

- - - - -   - - - - -   - - - - -

Put the pictures in order. Write 1, 2, and 3 to show the order.

Name _____

January

Sunday	Monday	Tuesday	Wednesday	Thursday	Friday	Saturday
	1		3		5	6
	8	9		11	12	
14		16	17	18		20
	22	23		25		27
	29	30		✕	✕	✕

Cut out the numbers on the next page. Glue each in the correct place on the calendar.

Chapter 10

Name _____

21	7	31	15

2	13	26	10

4	24	19	28

Cut out the numbers and glue on the calendar.

Chapter 10

Name

_____ o'clock

_____ o'clock

_____ o'clock

_____ o'clock

_____ o'clock

_____ o'clock

Write the number that tells the time.

Chapter 10

Name _____

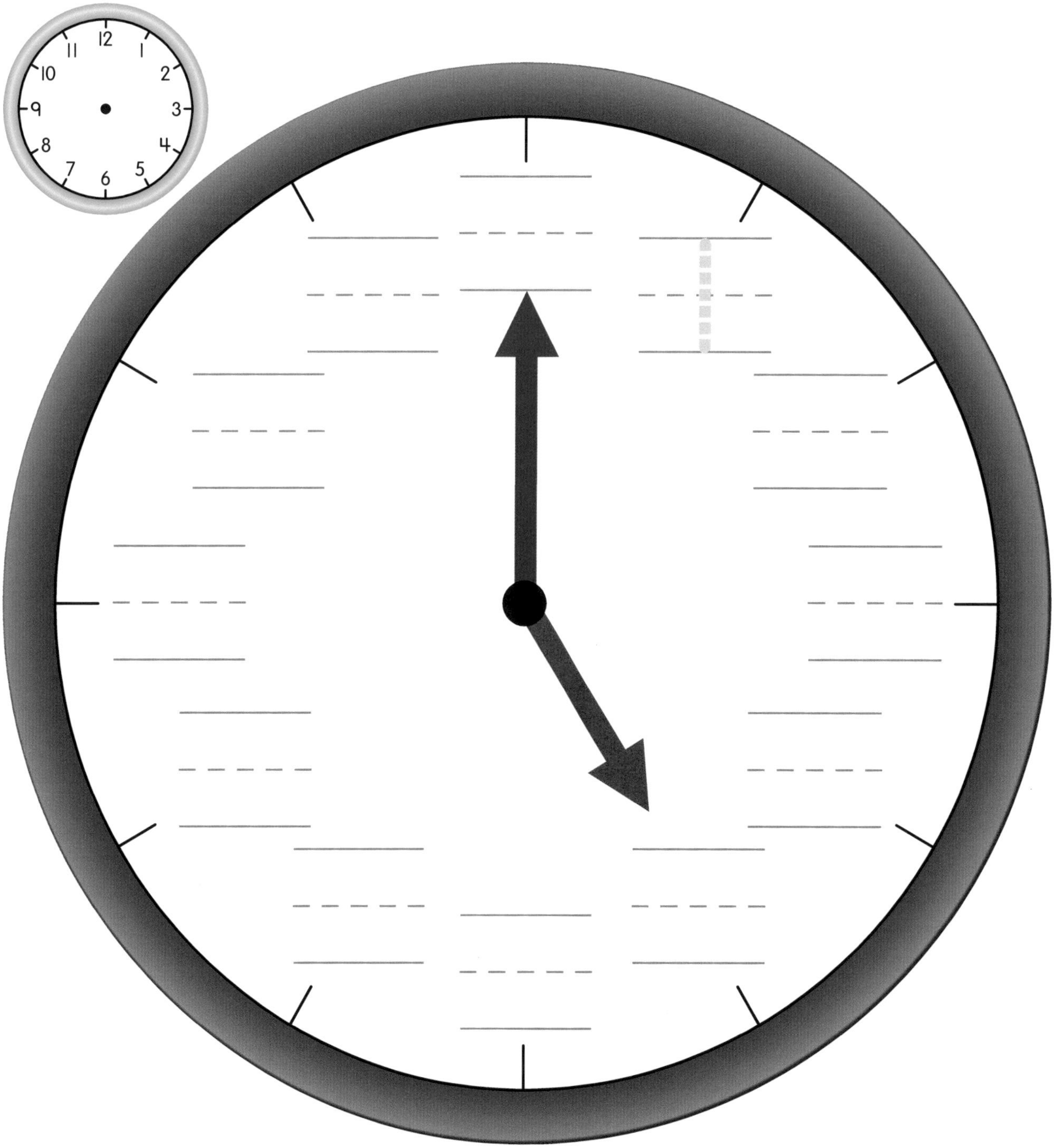

_____ o'clock

Write the numbers on the clock. Write the number that tells the time.

Name

7:00

3:00

10:00

12:00

5:00

9:00

© Calvert School

Read the digital clock. Match to the analog clock.

Chapter 10   (two hundred fifty-five) **255**

Name _____

 3 o'clock

 3:30 or half past the hour

Long hand (the minute hand) is on 6.

Short hand (the hour hand) is between 3 and 4.

_____ o'clock

_____ thirty

_____ o'clock

_____ thirty

_____ o'clock

_____ thirty

Write the number that tells the time.

Name

4 o'clock

12 o'clock

6 o'clock

10 o'clock

8 o'clock

7 o'clock

1 o'clock

5 o'clock

11 o'clock

Draw the hands to show the time.

Name _____

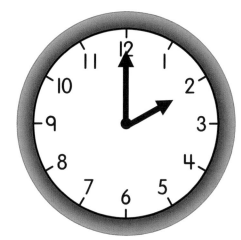

"X" the clocks whose numbers are incorrect.

Name

_ _ _ _ _ _ ¢

_ _ _ _ _ _ ¢

_ _ _ _ _ _ ¢

_ _ _ _ _ _ ¢

_ _ _ _ _ _ ¢

_ _ _ _ _ _ ¢

Count the coins in each box. Write the amount.

Chapter 10

Name _____

_____ ¢

_____ ¢

_____ ¢

_____ ¢

_____ ¢

_____ ¢

Count the coins in each box. Write the amount.

Name

 ● ●

 ● ●

 ● ●

 ● ●

 ● ● 6¢

© Calvert School

Match the coins and the amounts.

Name _____

Put an X on each box that is worth 10¢.

Name _____

 ● ●

 ● ●

 ● ●

 ● ●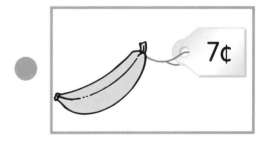

 ● ●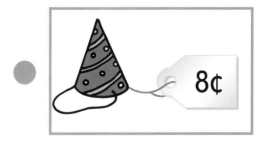

Draw lines to match the groups of coins to the prices.

Name _____

Put an X on the quarters. Circle the dimes.

Name

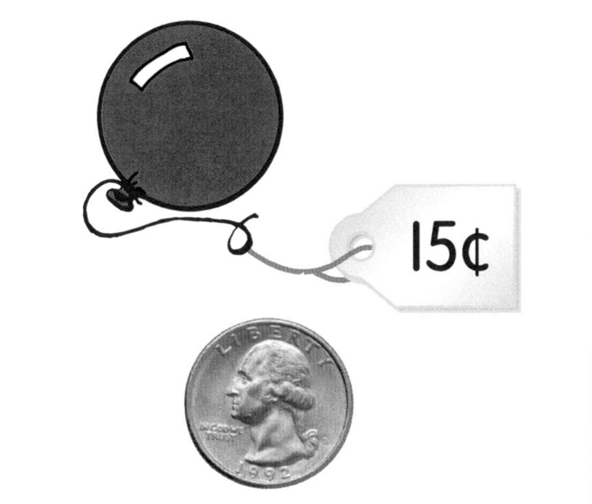

15¢

26¢

8¢

6¢

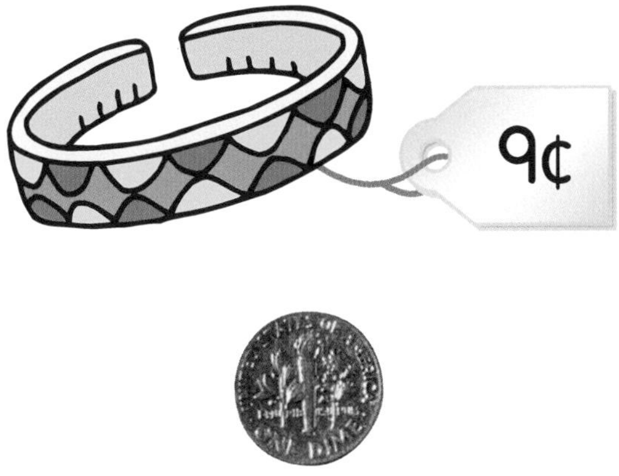

9¢

8¢

Circle the picture if there is enough money to buy the object in each box.

Chapter 10

Name

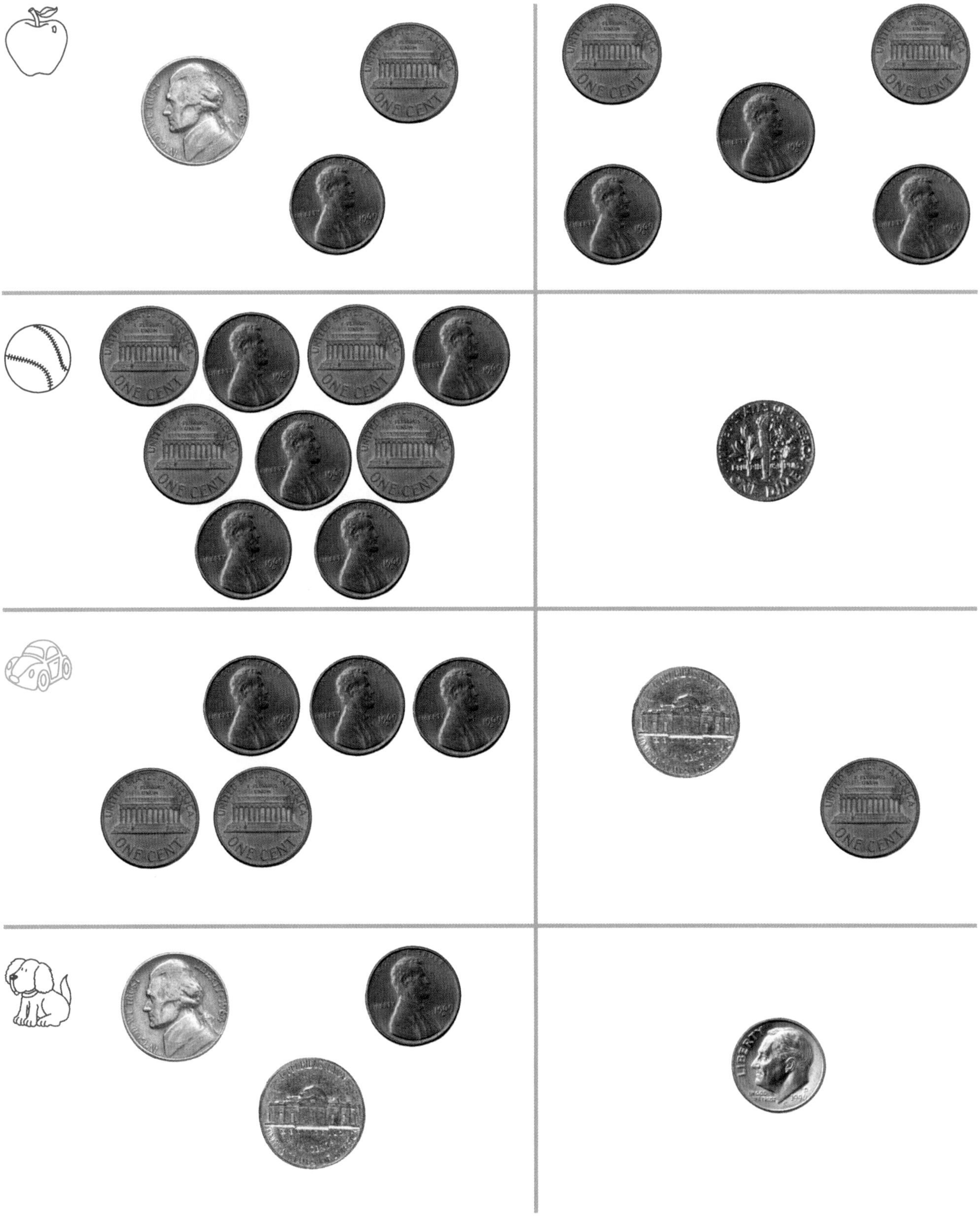

Count the coins. Put an X on the group in each row
that is less.

Name

Count the coins. Circle the group in each row that is more.

Name

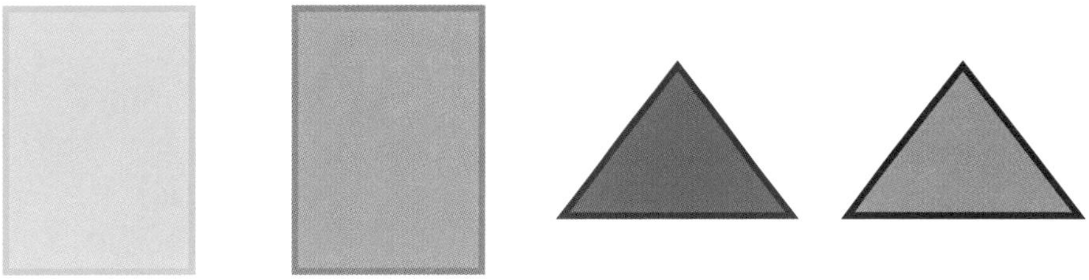

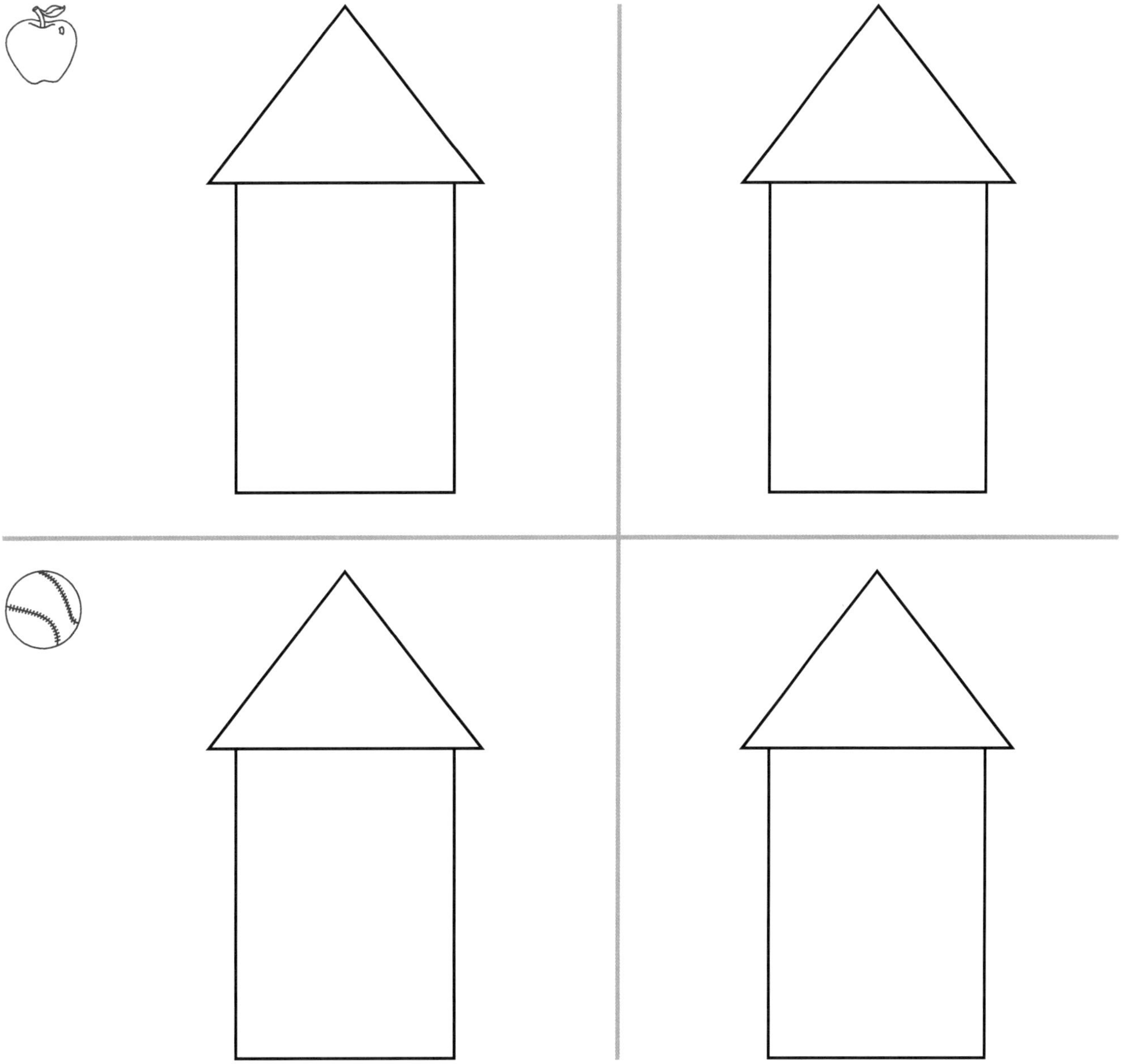

Color to show different ways the houses
can be put together.

Name

 Circle the 1st dog with a blue crayon.

 Put an X on the 3rd duck with a red crayon.

 Draw a triangle around the 5th cat with a green crayon.

 Color the 4th horse with an orange crayon.

　　　　　　　　　　　　　　　Chapter 11

Name

 Put a red X on the 3rd lion.

 Put a blue circle around the 2nd alligator.

 Put a green triangle around the 5th zebra.

 Put a blue X on the 1st fish.

 Put a red circle around the 4th cat.

Name

Guess how many birds there are. Write the number on the red lines.
Then count the birds. Write that number on the blue lines.

Name

People	Number of pieces of pizza	Number of ice cream cones
you		
your friends		
your family		
your neighbors		
your town		

Estimate how many pieces of pizza and how many ice cream cones each group would need.

Name

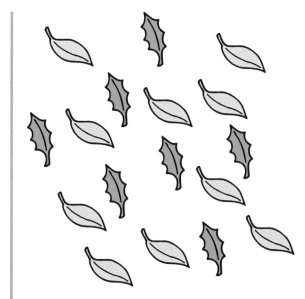

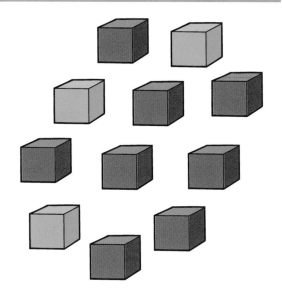

Circle the object you would more likely pick if the objects were in a bag.

Chapter 11

Name

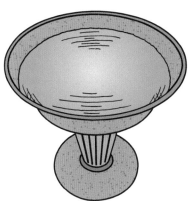

Look at the first picture. Then circle the picture that shows
what will happen next.

Name _____

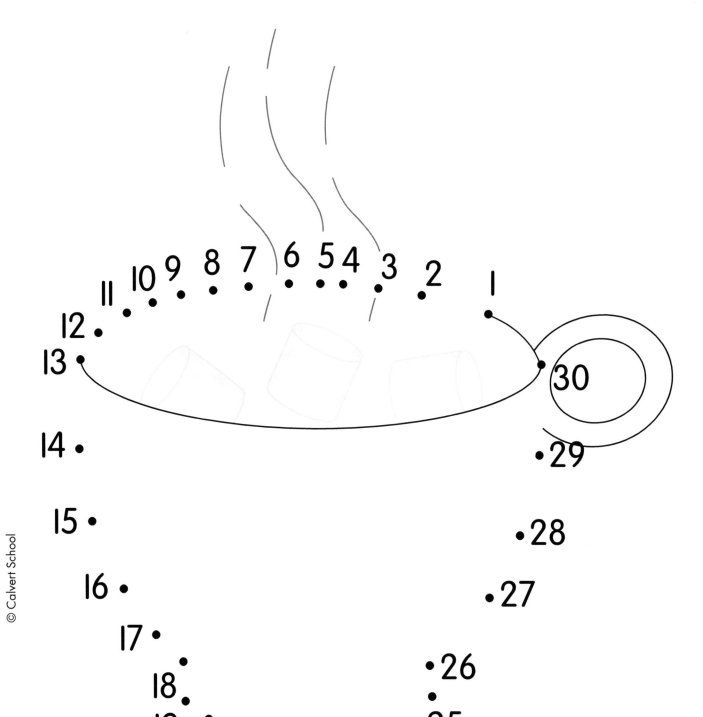

© Calvert School

Connect the dots to complete the picture.

Chapter 11

Name _____

Winter Night Graph

Number of Objects				
7				
6				
5				
4				
3				
2				
1				

Objects

Count the number of each object in the picture. Fill in the graph.

Name _____

$$5 + 1$$

$$3 + 3$$

$$2 + 3$$

$$4 + 1$$

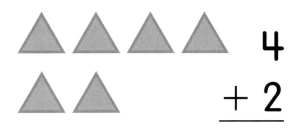

$$4 + 2$$

$$1 + 2$$

Add. Write your answers.

Name

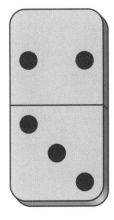

$$\begin{array}{r} 2 \\ + 3 \\ \hline \end{array}$$

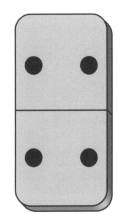

$$\begin{array}{r} 2 \\ + 2 \\ \hline \end{array}$$

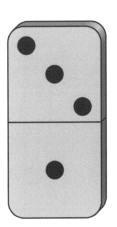

$$\begin{array}{r} 3 \\ + 1 \\ \hline \end{array}$$

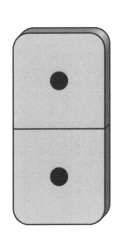

$$\begin{array}{r} 1 \\ + 1 \\ \hline \end{array}$$

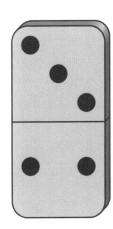

$$\begin{array}{r} 3 \\ + 2 \\ \hline \end{array}$$

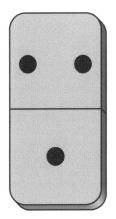

$$\begin{array}{r} 2 \\ + 1 \\ \hline \end{array}$$

Add. Write your answers.

Chapter 11

Name

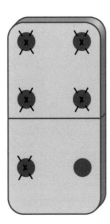

$$\begin{array}{r} 6 \\ -\ 5 \\ \hline \end{array}$$

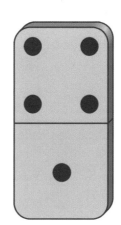

$$\begin{array}{r} 5 \\ -\ 2 \\ \hline \end{array}$$

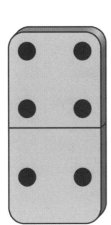

$$\begin{array}{r} 6 \\ -\ 2 \\ \hline \end{array}$$

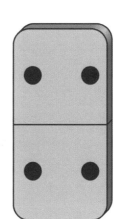

$$\begin{array}{r} 4 \\ -\ 3 \\ \hline \end{array}$$

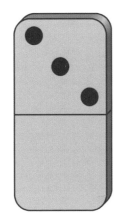

$$\begin{array}{r} 3 \\ -\ 2 \\ \hline \end{array}$$

$$\begin{array}{r} 1 \\ -\ 1 \\ \hline \end{array}$$

Cross out the number of dots being subtracted. Write your answers.